Wyvill James Christy

A Practical Treatise on the Joints Made and Used by Builders

Wyvill James Christy

A Practical Treatise on the Joints Made and Used by Builders

ISBN/EAN: 9783743687271

Printed in Europe, USA, Canada, Australia, Japan

Cover: Foto ©berggeist007 / pixelio.de

More available books at **www.hansebooks.com**

A PRACTICAL TREATISE

ON THE

JOINTS MADE AND USED BY BUILDERS

IN THE CONSTRUCTION OF VARIOUS KINDS OF ENGINEERING
AND ARCHITECTURAL WORKS, WITH ESPECIAL REFERENCE
TO THOSE WROUGHT BY ARTIFICERS IN ERECTING
AND FINISHING HABITABLE STRUCTURES

BY

WYVILL J. CHRISTY
ARCHITECT AND SURVEYOR

With upwards of One Hundred and Sixty Engravings on Wood

LONDON
CROSBY LOCKWOOD AND CO.
7, STATIONERS' HALL COURT, LUDGATE HILL
1882

LONDON:
PRINTED BY J. S. VIRTUE AND CO., LIMITED,
CITY ROAD.

PREFACE.

THE present work is compiled with the object of its forming a suitable companion to the useful Treatises on Architecture and Engineering included in this series, and which have proved so popular.

In offering it to the notice of the building trades, and to that of the professions which dominate them, it is hoped that some practical and useful result will accrue from thus collecting the various Joints made by the artificers of each trade in a more or less systematic and novel form. Admitting of easy reference, such an arrangement gives a synoptical insight into some of the most important operations of all the trades, whilst it may possibly facilitate the director or designer of work in choosing good forms of joints, and in securing their workmanlike execution.

It may be taken for granted that, as a rule, different parts of engineering and architectural works hold well together, for there would not be so much immunity from accidents from falling bodies were not architects habituated to wait submissively on the laws of stability—repose, equipoise, and tie—nor would the severe and studious labours of engineers to make bond, bolt, and rivet subserve them to the best advantage be either so creditable or profitable did not a similar exemption prevail throughout their works and triumphs.

It does not follow from this, however, that joints are always made as they should be. Indeed, there is a peculiar aspect which they bear towards the materials of a structure that points to the contrary, and which must not be overlooked in a work of this nature. Being formed by the juxtaposition of every block, brick, timber, slate, or piece of material, howsoever united, joints are wellnigh ubiquitous, and it is easy to see that they form a large part of the workmanship of an edifice, and that, in fact, good workmanship must in all trades be synonymous with good jointing. It is, moreover, equally clear that there must be a frequent temptation and tendency to introduce into a building either indifferent materials or indifferent workmanship, or a reprehensible blending of the two.

The question then is whether, having regard to security against accident in engineering works, or to the wholesomeness and durability of habitable structures and house property in general, it is not the better of two evils to have inferior materials put together in a workmanlike manner, that is, properly jointed, rather than to build in materials of a superior class by means of hastily executed and more or less imperfect joints.

Considering that joints are admittedly the weakest parts of structures, it follows, as a matter of course, that the former is by far the best preventive and alternative; for, to say nothing of impending danger from joints weakened by the inroads of rust, or in jeopardy from the chance passage of too heavy loads, it is notorious enough that gales annually cause much brickwork, &c., to part at the joints, whilst sanitarians boldly ascribe to defective pipe joints no trivial amount of preventible ills. Again, how can a house or other human habitation be well and substantially built, or, in other words, continue sound proof, damp proof, and wind proof, unless the walls keep dry and plumb, the floor and roof timbers sit immovably upon their bearings, and the whole roof-covering remains staunch and impervious after the pelting downfall or searching storm? All this is the legitimate outcome of sound jointing, and so is what is needful to the same desirable end as performed in finishing a building. Doors, windows, and floors will all be draughty unless well fitted and hung, or well laid and compacted, whilst the plasterer must leave no chinks nor the gasfitter leaks. Though good materials may show a better face, the magic influence of skilled labour, properly applied, irresistibly vindicates its supremacy, and turning middling ones to the best account, aggregates them into a noble structure, both safe and firm, or into a healthful and desirable abode.

In the compilation of this treatise no inconsiderable observation and experience have been supplemented by recourse to many standard works on the practice of building, without which aid few would have ventured on such a task. Whilst frankly acknowledging the obligation due to these and other kindred publications, the author would fain dispel the pardonable idea that their accumulated information leaves little room for further dissertation on the subject of jointing, for the following pages even present the great family of joints in a somewhat new aspect, augmented by fresh relatives and connections gathered out of old parts, processes, and contrivances. Thus, for instance, the well-known terms *torching* and *flashing* represent respectively a process and an object, yet characteristic joints accompany their adoption, and, analogically, it is as correct to speak of a torched or flashed joint as it is to denominate the outcome of mortising a mortised joint. At all events, in both cases a weather-tight joint is the result. Hence it has been considered the wisest course to incorporate these and others of corresponding extraction rather than exclude from a book wholly devoted to the explanation of joints any forms thereof having positive claims to be classed as such.

It only remains to append a list of the chief Works consulted in the preparation of this volume :—

Cresy's Encyclopædia of Civil Engineering.—Spon's Dictionary of Engineering.—Rankine's Civil Engineering.—The Student's Practical Guide to Measuring and Valuing.—Barlow on the Strength of Materials.—Newland's Carpenter and Joiner's Assistant.—Barlow's Tredgold's Carpentry.—Hellyer's Plumber and Sanitary Houses.—Building Construction, published by Rivington.—Latham's Sanitary Engineering.—Donaldson's Specifications.—Anderson's Strength of Materials. —Building Construction, by Burn.—Reid's Concrete.—Davy's Artificial Foundations.—Gwilt's Encyclopædia of Architecture.—Bartholomew's Specifications.—Hutchinson's Girder Making and Bridge Building.— Campin's Iron Roofs.—Humber's Handy Book of Strains.—Matheson's Works in Iron.—Knight's Dictionary of Mechanics.—The Professional Periodicals.—Nicholson's Practical Carpentry and Joinery.—Nicholson's Practical Masonry, Bricklaying, and Plastering.—Buck's Oblique Bridges.—The Dictionary of Architecture.

And the following works in Weale's Rudimentary Series, viz. :—

Civil Engineering.—Dictionary of Terms.—Drainage of Towns and Buildings.—Drainage of Districts and Lands.—Gas Works.—The Art of Building.—Masonry and Stonecutting.—Foundations and Concrete Work.—Practical Bricklaying.—Plumbing.—Roads and Streets.— Carpentry and Joinery.—Iron Bridges, Girders, Roofs, &c.

W. J. C.

CONTENTS.

INTRODUCTORY 1

SECTION I.—JOINTS USED IN DRAINAGE.

	PAGE		PAGE
Asphalte Joint	6	Half Socket Joint	8
Bed or Bedding Joint . .	6	Lip Joint	8
Butt Joint	7	Pipe ,, ...	8
Cement ,, ...	7	Socket ,, ...	9
Clay ,, ...	7	Stanford Joint	9
Collar ,, ...	8	Tight ,, ...	10

SECTION II.—BRICKLAYERS' JOINTS.

Angular Joint	11	Flat Joint	20
Arch ,, ...	11	Flat Joint Jointed	20
Asphalte ,, ...	12	Flat Struck and Jointed	
Back ,, ...	12	Joint	20
Bastard Tuck Pointed Joint	12	Flush Joint	21
Bead Joint	12	Flushed ,, ...	21
Bed or Bedding Joint . .	12	Flush Jointed Joint . . .	21
Bevel or Bevelled ,, . .	14	Full Joint	21
Birdsmouth	14	Grouted Joint	21
Black or Blue Pointed Joint	14	Hacked-out Joint	22
Bond or Bonding Joint . .	14	Heading ,, ...	22
Broken Joint	16	High ,, ...	22
Cement ,, ...	16	Hollow ,, ...	22
Chimney Pot Joint . . .	16	Jointed or Jointer Joint . .	23
Close Joint	17	Keyed Joint	23
Coarse ,, ...	17	Lap ,, ...	23
Collar ,, ...	18	Lip ,, ...	23
Coursing Joint	18	Loose ,, ...	23
Cross ,, ...	18	Mitre ,, ...	23
Cut and Struck Joint . . .	19	Mortar ,, ...	24
Diagonal Joint	19	Parallel ,, ...	25
Dipped ,, ...	19	Pat ,, ...	25
Dovetail ,, ...	19	Perpendicular Joint . . .	25
Drawn ,, ...	19	Piecing Joint	25
Face ,, ...	20	Pointed ,, ...	26
False ,, ...	20	Putty ,, ...	26
Filleted ,, ...	20	Radiating ,, ...	26
Fine ,, ...	20	Raised ,, ...	26

x   CONTENTS.

	PAGE		PAGE
Raked Joint	26	Thick Joint	28
Recessed ,,	27	Thin ,,	29
Rough ,,	27	Tie ,,	30
Rubbed ,,	27	Toothed ,,	30
Saddle ,,	27	Transverse Joint	30
Shift ,,	27	Tuck Pointed Joint	30
Slip ,,	27	Upright Joint	31
Spring ,,	27	Vertical ,,	31
Straight ,,	27	Wall ,,	31
Struck ,,	27	Weather ,,	32
Sunk ,,	28	White ,,	32
Terra Cotta Joint	28	Wood ,,	32

SECTION III.—MASONS' JOINTS.

	PAGE		PAGE
Angular Joint	34	Fine Tooled Joint	48
Arch ,,	34	Flushed Joint	48
Asphalte ,,	35	Full ,,	49
Axed ,,	35	Grooved ,,	49
Back ,,	35	Grooved and Tongued Joint	50
Back Rebate Joint	35	Grouted Joint	50
Bag Joint	36	Heading ,,	50
Bed or Bedding Joint	36	Hollow ,,	51
Birdsmouth	37	Horizontal ,,	51
Blown Joint	38	Housed ,,	51
Bolted ,,	38	Joggle ,,	51
Bond or Bonding Joint	39	Lap or Lapped Joint	52
Broken Joint	41	Lead or Leaded ,,	52
Butt ,,	41	Mason's Joint	53
Caulked ,,	42	Mastic ,,	53
Cement ,,	42	Mitre ,,	53
Chamfered Joint	43	Mortar ,,	53
Checked ,,	43	Mortise Joint	54
Circular ,,	43	Mortise and Tenon Joint	54
Clamped ,,	43	Notch or Notched Joint	54
Close ,,	43	Open Joint	55
Coarse ,,	44	Perpendicular Joint	55
Coursing ,,	45	Pinned Joint	55
Cramp or Cramped Joint	45	Plugged ,,	56
Cross Joint	46	Pointed ,,	56
Dovetail Joint	46	Putty ,,	57
Dowelled Joint	46	Radiating Joint	57
Elbow ,,	47	Raglet ,,	57
End ,,	47	Raised ,,	58
Face ,,	47	Raked ,,	58
Fair ,,	47	Rebated ,,	58
Fair Axed ,,	47	Recessed ,,	58
False ,,	48	Rough-tooled Joint	59
Filleted ,,	48	Rubbed Joint	59
Fine ,,	48	Run ,,	59

CONTENTS.

	PAGE		PAGE
Rustic or Rusticated Joint	60	Thick Joint	63
Saddle Joint	60	Thin ,,	63
Sawn ,,	61	Tie ,,	63
Scribed ,,	61	Tooled ,,	63
Shift ,,	61	Transverse Joint	64
Side ,,	61	True Joint	64
Smooth ,,	62	Underpinned Joint	64
Square ,,	62	Upright Joint	64
Straight ,,	62	V-Joint	64
Strap or Strapped Joint	62	Vertical Joint	65
Sunk Joint	62	Water ,,	65
Tabled ,,	62	Weather ,,	65

SECTION IV.—TILERS' JOINTS.

	PAGE		PAGE
Back Joint	67	Heading Joint	69
Bed or Bedding Joint	67	Key ,,	69
Bond or Bonding Joint	68	Lap ,,	69
Broken Joint	68	Mitre ,,	69
Cement ,,	68	Mortar ,,	69
Dowelled ,,	68	Pointed ,,	69
Filleted ,,	68	Weather ,,	69
Flanched ,,	68		

SECTION V.—SLATERS' JOINTS.

	PAGE		PAGE
Back Joint	71	Lap Joint	73
Bed or Bedding Joint	71	Mastic Joint	74
Bond or Bonding Joint	72	Mortar ,,	74
Broken Joint	72	Open ,,	74
Cement ,,	72	Painted ,,	74
Cistern ,,	72	Pointed ,,	75
Close ,,	72	Rebated ,,	75
Close Cut Joint	72	Shouldered Joint	75
Close Cut and Mitred Joint	72	Side Joint	75
Filleted Joint	73	Torched Joint	76
Heading ,,	73	Vertical ,,	76
Key ,,	73	Weather ,,	76

SECTION VI.—CARPENTERS' JOINTS.

	PAGE		PAGE
Abutting Joint	78	Brace Joint	84
Adjustable ,,	78	Bridle ,,	84
Angle ,,	79	Butt or Butting Joint	84
Angular ,,	79	Carpenter's Boast	85
Angular Grooved and Tongued Joint	79	Caulked Joint	85
		Chase Mortise Joint	86
Banded Joint	79	Checked Joint	86
Bed or Bedding Joint	80	Circular ,,	86
Bevel or Bevelled ,,	80	Close ,,	86
Bevelled Shoulder ,,	81	Cocked ,,	87
Birdsmouth	81	Cogged ,,	87
Bolted Joint	81	Corked ,,	87

CONTENTS.

	PAGE			PAGE
. . . .	87	Open Joint		98
. . . .	87	Piecing ,, ...		98
Joint . .	87	Pinned ,, ...		98
. . . .	87	Plain or Plane Joint . . .		100
Halved		Post and Beam ,, . . .		100
. . . .	88	Pulley Mortise ,, . . .		100
Joint . .	88	Radiating ,, . . .		100
,, . .	88	Rebated ,, . . .		100
or Foxtail		Rounded ,, . . .		101
. . . .	89	Scarf or Scarfed ,, . . .		101
. . . .	89	Screw Shackle ,, . . .		102
gued Joint	89	Serrated ,, . . .		102
. . . .	90	Ship Lapping ,, . . .		103
. . . .	90	Shouldered ,, . . .		103
d Joint . .	90	Socket ,, . . .		103
. . . .	90	Spiked ,, . . .		104
. . . .	91	Splayed ,, . . .		104
tailed Joint	91	Splice ,, . . .		104
. . . .	91	Square ,, . . .		104
. . . .	91	Stirrup ,, . . .		105
. . . .	91	Strap or Strapped Joint . .		105
. . . .	92	Strap and Bolt ,, . .		106
. . . .	92	Strap and Cotter ,, . .		107
int . . .	92	Strut Joint		107
. . . .	93	Swallow-tail Joint		107
oint . .	93	Tabled Joint		108
,, . . .	94	Tabled Indented and Keyed		
,, . . .	94	Joint		108
. . . .	95	Tie Joint		108
n Joint .	95	Toothed Joint		109
,, . .	97	Transverse ,, ...		109
,, . .	97	Water-tight Joint		109
. . . .	97	Wedged ,, ...		109

SECTION VII.—SMITHS' JOINTS.

	PAGE		PAGE
. . . .	111	Caulked Joint	121
. . . .	111	Cement ,, ...	122
. . . .	112	Chain Riveted Joint . . .	122
. . . .	113	Chamfered ,, . .	123
. . . .	113	Chipped and Filed Joint . .	123
. . . .	114	Circular Joint	123
. . . .	114	Clamped ,, ...	124
. . . .	114	Clip ,, ...	124
oint . . .	115	Close ,, ...	124
. . . .	117	Collar ,, ...	124
. . . .	119	Corked ,, ...	125
. . . .	119	Cottered ,, ...	125
. . . .	119	Coupling ,, ...	126
. . . .	120	Cover ,, ...	127
. . . .	121	Cramp or Cramped Joint .	127

CONTENTS.

	PAGE		
Diagonal Joint	127	Radial Joint	
Double Nut Joint	127	Radiating ,,	
Double Riveted Joint	127	Rebated ,,	
Dovetail Joint	128	Red Lead ,,	
Dowelled ,,	128	Riveted ,,	
Drilled ,,	128	Run Joint	
Elbow ,,	129	Rust ,,	
Expansion ,,	129	S ,,	
Eye ,,	130	Saddle ,,	
Faced ,,	131	Sand ,,	
Faucet ,,	131	Screw ,,	
Ferrule ,,	131	Screw Shackle Joint	
Fish or Fished Joint	132	Shackle Joint	
Flange Joint	132	Shoe Joint	
Folded Angle Joint	133	Single Riveted Joint	
Grooved Joint	134	Socket Joint	
Grouped Riveted Joint	134	Soldered ,,	
Hinge Joint	134	Spigot and Faucet Joint	
Hook ,,	134	Spigot and Socket ,,	
Horizontal Joint	135	Strut Joint	
India-rubber Joint	135	Sulphur ,,	
Iron Cement ,,	135	T ,,	
Joggle Joint	136	Thimble ,,	
Jump ,,	136	Through and Through Joint	
Key ,,	136	Tie Joint	
Lap or Lapped Joint	136	Tight Joint	
Lead or Leaded ,,	136	Triple Riveted Joint	
Lightning Conductor Joint	137	True Joint	
Longitudinal Joint	137	Turned and Bored Joint	
Mitre Joint	137	Union Joint	
Mortise and Tenon Joint	137	Universal Joint	
Pin Joint	138	Vertical ,,	
Pipe ,,	138	Water-tight Joint	
Pivot or Pivoted Joint	139	Wedged ,,	
Planed Joint	140	Welded ,,	
Punched ,,	140	Zigzag Riveted ,,	

SECTION VIII.—JOINERS' JOINTS.

Angle Joint	157	Close Joint	
Angular Joint	157	Cramp or Cramped Joint	
Angular Grooved and Tongued Joint	157	Cross Grooved and Tongued Joint	
Beaded Joint	157	Double Lapped Joint	
Broken ,,	158	Double Quirk Bead Joint	
Butt ,,	158	Double Tongued Joint	
Chamfered ,,	158	Dovetail Joint	
Checked ,,	158	Dowelled ,,	
Circular ,,	158	Feather or Feather Tongued Joint	
Cistern ,,	158		
Clamped ,,	159	Feather Wedged Joint	

	PAGE		PAGE
Fillistered Joint	161	Mortise and Tenon Joint	169
Folding „	161	Nailed Joint	169
Fox Wedged „	161	Notched Joint	169
Framed „	161	Open „	170
Franked „	161	Piecing „	170
Glued or Glued up Joint	162	Pinned „	170
Glued and Blocked „	163	Plain or Plane Joint	170
Grooved and Feathered Joint	163	Ploughed and Feathered Joint	171
Grooved and Rebated „	163	Ploughed and Tongued Joint	171
Grooved and Tongued „	163	Plugged Joint	171
Grooved Tongued and Beaded Joint	164	Rabbeted „	171
Grooved Tongued and Mitred Joint	164	Radius „	171
		Rebated „	171
Half Mortise and Tenon Joint	164	Rebated and Beaded Joint	172
Hammer Headed Key Joint	164	Rebated and Filleted „	172
Handrail Joint	164	Rebated and Mitred „	172
Heading Joint	165	Rebated Grooved and Tongued Joint	172
Hinge „	165	Rounded Joint	173
Hook „	165	Rule „	173
Housed „	165	Running „	173
India-rubber Joint	166	Scarf or Scarfed Joint	173
Key Joint	166	Screw Joint	174
Keyed „	166	Scribed „	174
Keyed Mitre Joint	166	Seam „	175
Lap or Lapped „	166	Secret Dovetail Joint	175
Lap or Lapped Dovetail Joint	167	Secret Nailed „	175
		Shoe Joint	175
Lapped and Mitred Dovetail Joint	167	Shutting Joint	176
Lapped and Tongued Mitre Joint	167	Sliding Joint	176
		Slip „	177
Lapped Mitre Joint	167	Slip Feather Joint	177
Lateral Joint	167	Splice Joint	177
Longitudinal Joint	167	Sprigged Joint	177
Mitre Joint	167	Spring „	177
Mitre and Butt Joint	168	Square „	178
Mitred and Cross Tongued Joint	168	Square Edged Joint	178
		Straight Joint	178
Mitre Clamped Joint	168	V-Joint	178
Mitred Dovetail „	168	Water-tight Joint	178
Mitre Joint Keyed	169	Weather Joint	178
		Wedged „	179

SECTION IX.—PLASTERERS' JOINTS.

Angular Joint	181	Key Joint	181
Broken „	181	Mitre „	182
Butt „	181	Tongued Joint	182
False „	181	Toothed „	182
Grooved „	181		

CONTENTS. XV

SECTION X.—PLUMBERS' JOINTS.

	PAGE		PAGE
Astragal Joint	184	Longitudinal Joint	196
Autogenous Soldered Joint	184	Mitre Joint	196
Bed or Bedding Joint	185	Overcast Joint	196
Block Joint	185	Overcast Ribbon Joint	196
Blown „	186	Overlapping Joint	196
Bottle Nose Drip Joint	186	Pipe Joint	197
Branch Joint	187	Plugged Joint	197
Burnt in „	187	Putty „	198
Butt „	188	Raglet „	198
Caulked „	188	Ribbon „	199
Cement „	188	Roll „	199
Cistern „	188	Rolled „	200
Cone „	188	Round „	200
Copper-bit Joint	189	Run Joint	200
Dog's Ear Joint	191	Screw „	202
Double Cone Joint	191	Seam „	202
Drawn Joint	191	Short „	203
Drip „	191	Slip „	203
Elastic „	192	Slip Socket Joint	204
Elbow „	192	Socket Joint	204
Expansion Joint	192	Soldered „	205
Faucet „	193	Striped „	206
Flange „	193	Swivel „	206
Flashed Joint	194	T-Joint	206
Flexible „	194	Taft „	206
Float or Flow Joint	194	Tight Joint	207
Flush Soldered „	194	Underhand Joint	207
Heading Joint	195	Union Joint	208
Hollow Roll Joint	195	Upright „	208
Knee Joint	195	Wedged „	209
Lap or Lapped Joint	195	Welded „	209
Lead or Leaded „	195	Wiped „	209
Long Joint	196		

SECTION XI.—ZINCWORKERS' JOINTS.

Capped Joint	213	Mitre Joint	215
Clip „	214	Roll Joint	215
Drip „	214	S-Joint	215
Flashed „	214	Saddle Joint	215
Fold „	214	Screw „	215
Folded „	214	Soldered Joint	215
Folded Angle Joint	214	Wedged „	216
Heading Joint	214	Welted „	216
Lap or Lapped Joint	215	Zinc Roll Cap Joint	216

SECTION XII.—COPPERSMITHS' JOINTS.

	PAGE		PAGE
Brazed Joint	217	Folded Joint	218
Capped „	218	Lap or Lapped Joint	218
Copper Roll Cap Joint	218	Lightning Conductor Joint	218
Cramp or Cramped Joint	218	Roll Joint	219
Fold Joint	218		

SECTION XIII.—GLAZIERS' JOINTS.

Beaded Joint	220	Open Joint	222
Bed „	220	Painted „	222
Butt „	220	Putty „	222
Cement „	220	Puttyless Joint	223
Filleted „	221	S-Joint	223
India-rubber Joint	221	Sprigged Joint	223
Lap or Lapped „	221	Thermo-plastic Putty Joint	223
Lead or Leaded „	221	Wash-leather Joint	224
Overlapping „	222		

SECTION XIV.—GASFITTERS' JOINTS.

Ball and Socket Joint	225	Sliding Joint	229
Barrel Union „	225	Socket „	229
Blown Joint	226	Soldered „	229
Blowpipe Joint	226	Stiff „	229
Copper-bit „	227	Swing „	229
Cup and Ball Joint	228	Swivel „	230
Elbow Joint	228	T-Joint	230
Hydraulic Joint	228	Tight Joint	230
Knee Joint	228	Union „	232
Pipe „	228	Universal Joint	232
Saddle „	228	Water Joint	232
Screw „	229		

SECTION XV.—PAPERHANGERS' JOINTS.

Butt Joint	233	Mitre Joint	233
Close „	233	Pencil „	233
Lap „	233	Rolled „	233

SECTION XVI.—PAVIORS' JOINTS.

Asphalte Joint	234	Gravel Joint	235
Bed or Bedding Joint	234	Grouted „	236
Close Joint	235	Racked „	236

INDEX 237

ON
THE JOINTS MADE AND USED BY BUILDERS.

INTRODUCTORY.

A JOINT is the place where two different portions or pieces of a structure meet or unite, and it consequently forms the division or union between them, as well as a constituent factor of the structure it helps to consolidate. A joint is either simple or compound, fixed or movable, and however perfect it may be it is almost invariably a source of weakness. In this light the monolithic description of concrete is superior to masonry; but, as an opposite instance, the superiority of a beam sawn asunder down the middle, reversed, and bolted together again, over a jointless one, is worthy of note. As a general rule all parts of a joint should be equally strong, the strength of the entire joint equal if possible to that of the parts joined, and its form, when used in framing or other analogous combination, such as to direct the pressures, as far as skill will allow, along the axes of the component pieces. Moreover, a joint should never be of a complicated nature, but its parts, reduced to the smallest practicable number, ought to be so devised as to resist in the most scientific manner the particular kind of stress that each will be tried by, whilst as a whole it satisfies the object of its formation through economic construction, fitness, and soundness.

By a simple joint is meant one consisting of nothing beyond the contact of meeting surfaces, such as an abutting joint in carpentry, without fastenings of any kind—though it may otherwise ill-devisedly be of an elaborate and intricate character; and by a compound joint is meant one in which a cementing medium, or some kind of attachment, is used to assist weight or pressure in keeping the contiguous parts from moving or racking—the first step towards separation and ruin. A mortar joint forms a common illustration of the latter class, for the mortar not only binds the bricks together, but before setting affords a conveniently yielding bed to enable each brick to take its bearing, so to speak, and thus to be in its best position to resist external forces. Such a joint is necessarily a compound one, since the mortar joint proper between any two bricks is compounded of the two joints formed by the adhesion of the intervening mortar to the surface of both bricks. Either a compound or a simple joint may be fixed or movable, the sliding joint in joinery affording a familiar instance of the simple movable variety. In construction generally joints are for the most part compound and fixed as well, but whether in engineering works or in the carcase or finishing of a building, the simple fixed description is of common occurrence; as instanced, for example, in the bed joint at the bearing end of a flitch beam seated on a stone template, and in the key joint between the plasterer's rendering and the wall's rough surface. The movable sort, which includes hinged, expansion, adjustable and lifting joints, appears occasionally in carpenters', joiners', smiths', and plumbers' work.

A compound joint consists, as has been said, of something besides the meeting or united surfaces, and this may be either a cementing medium such as mortar, cement, asphalte, glue, white lead, solder, sulphur, lead, cast zinc, putty, clay, &c., or else a fastening or connection consisting of one or more of the following:—hoops, straps with or without gibs and cotters, cast or wrought iron shoes and sockets, screw bolts, wedges, dowels, cramps, bolts, pins, keys, treenails, rivets, plates, nails, spikes, sprigs, screws, &c. Sometimes one or other of these may be present in conjunction with a cementitious material, but in most instances it is found advisable, owing to the disturbing influence of settlement, shrinkage, expansion, impact, &c., not to trust alone to superincumbent pressure or the operation of gravity for the maintenance of joints, but to fortify simple or interlocking contact by the extraneous aid of a third substance.

And here it may be observed that a joint is wrongly defined when it is said to be the space occupied by the substance between two pieces of material. The surfaces of course must also be taken into account, else how can joints be rightly named, as is often the case, after the particular manner in which they are wrought, the common dovetail, for instance, furnishing a well-known illustration? It is equally wrong to define a joint to be an interstice, for besides being frequently distinguished by the way in which they are filled, joints ought to be full and close, excepting when a little slackness is left purposely in framing and elsewhere to allow for the inevitable settlement when the load is felt. The evils arising from punching involve much loose riveting, not only in the booms of girders but likewise at the junctions of the cross-bracing; but this is accidental, and the reason why joints in wood are so often open is because careless setting out, bad fitting, and imperfect seasoning cause unequal bearing or an excess of shrinkage, and preclude the surfaces from closing or keeping close either when at rest or under the stress of pressure, or when tried by variations of temperature or the hygrometric condition of the air. Quirks, mouldings, and architraves in joinery are notoriously useful in concealing the interstice, or air-space, arising from the shrinkage of insufficiently seasoned stuff; which air-space reminds one of the spot on the dinner-table that ought to have been occupied by the haunch of venison, which not turning up in due course, left it to the spot in question simply to indicate where the joint ought to have been.

The nomenclature of joints is for the most part determined by their shape, character, position, and mode of execution or finish; from the particular kind of cementing material, or fastening, or connection employed for preventing severance; and likewise from the structure to which they belong. We thus have rebated joint, coarse joint, bed joint, wiped joint, flushed joint, mortar joint, pinned joint, strap joint, arch joint. It is rarely, however, that a name is derived from that of the materials united, though the term metal joints is generically used to denote the whole class of joints between metallic surfaces. Jointing and joining are synonymous with joint, but the latter is not frequently used, whilst the former is commonly employed for the sham joints in stucco to imitate masonry. Amongst plumbers connection often stands for joint.

Omitting altogether the consideration of those cases in which a heavy draw is made upon the designer's art to combine strength and elegance in complicated joints, as, for example, where the extremities of level, upright, and inclined members cluster together for mutual support, perhaps there is no other branch of labour in the whole round of a general builder's business, which indicates and tests his responsibility and capacity so much as the execution in a workmanlike manner of the particular species of work now under notice. Excepting with regard to his own individual trade he is quite left in this respect to the skill, honesty, or pleasure of the craftsmen in the others; and when the number and variety of the joints are considered which they have thus to make, as well as the difficulties and obstacles which often stand in the way of making them sound and durable, it seems almost incredible that anybody, however practical yet incompetent to make them, and at the same time lacking the knowledge of the theory of their construction, should presume to guarantee their fitness and reliability. And this is more particularly the case since bad joints do not of necessity hurriedly disclose themselves. On the contrary, it often happens that storm and tempest, frost and drought, have to be encountered before strain, vibration, or fluctuation of dimensions demonstrate their indifferent or even beggarly character. Thus the question arises, can work performed under such conditions and supervision, unchecked by the scientific and general knowledge of an architect or engineer, produce carcases and finishings or habitable dwellings that if sold are not usually sold as being superior in construction and finish to what they really are? And again, assuming this to be the case, does the unwary purchaser suffer most by ignoring a surveyor, or by the builder being his own architect? Leaving these inquiries for the inquisitive to probe, it remains to describe succinctly and alphabetically under each trade the joints by which diversified units and substances are aggregated into a perfect structure. Within the narrow compass, however, of a brief explanation, it will not be feasible to particularise all the modifications and applications of each sort of joint that have sprung from widespread use influenced by individual fancy, local custom, or the effects of the ever restless desire for change and novelty, which is as active amongst the building fraternity as elsewhere, though perhaps more under control, owing to the risk to life and property attachable to a departure from the well-beaten track.

SECTION I.

JOINTS USED IN DRAINAGE.

Asphalte Joint is a means adopted for jointing socketed sewer-pipes, and is said to possess besides the qualities of adhesiveness, elasticity, and imperviousness, the power of resisting penetration by the rootlets of trees. It is made by first of all caulking with two or three strands of tarred gasket and then filling up the remainder of the cavity left in the socket by pouring in asphalte rendered liquid by heat, a temporary belt of clay being passed round the edge of the socket to confine it till solidification ensues.

Bed or Bedding Joint occurs between the bricks, &c., of brick drains and sewers, and should in all cases be solidly made with Portland cement or hydraulic lime; the interstices between the bricks caused by the quickness of the curve being filled in at the invert, &c., with pieces of slate or tile. It also exists between the invert blocks of sewers and their concrete or clay or other foundation or outside lining when built in porous soils, the puddle lining girting the lower half of the sewer. It occurs likewise between a pipe and the substance on which it lies, and the greatest care is needed in order that drain-pipes may have a solid bearing and bear equally throughout their beds, which must be absolutely exempt from the danger of settlement. The same remarks are applicable to the stoneware invert blocks of brick sewers. Surrounding pipes with concrete thrown into the trench after they are jointed not only strengthens their sides against fracture by crushing but lessens the danger of subsidence. Large pipes sometimes require holes purposely cut in the floor of the trench to take the sockets. In building walls over drain-pipes the space of a few inches must be left to allow for the wall's settlement, and the same care is required in piercing old walls for new drains. After springing a line of pipes up into a curved form to insert a junction, the invert must be again truly straightened and the compactness of the foundation restored.

Butt Joint is formed between the joints of agricultural or surface, or subsoil drain-pipes, consisting of short cylindrical tubes, by merely butting them against one another, end to end. The joint, however, must be protected by slipping over it a perforated collar when laid in running sand or where the bottom is too loose to keep the pipes in line.

Cement Joint.—For jointing socketed drain-pipes Portland cement is used, both neat and mixed with an equal part of clean sharp sand; but besides being porous, and therefore not air-tight, it is accused, by the advocates of clay luting and others, of cracking under the slightest settlement, motion, or expansion of the pipe; and in some cases its use obliges the trench to be pumped dry, which is not the case with the Stanford joint.

Clay Joint.—Good tough clay, well puddled, or, in other words, good clay puddle, is supposed to be impervious to water. Puddling consists in thoroughly beating up or tempering the clay with water and mixing with it some chalk, gravel, or sand, to prevent it from cracking on becoming dry. An interstice stopped by this substance is a clay joint, and brick sewers built in porous soils are surrounded up to half their height with a lining of puddle. Sewer-pipes are sometimes united with carefully tempered clay, which is considered by some superior to Portland cement for such purpose. The socket of each pipe is well daubed with it before laying, and the inside of the joint is worked solid and well smoothed after the plain end of the next is inserted. It is not so liable to crack under expansion or settlement, though it has been strongly denounced because of its soft, yielding character, which not only renders it liable to be washed out either by the sewage from the inside or the subsoil water from the outside, but endangers the maintenance of a straight line of invert through the weight of earth above the pipe squeezing the clay out of the lower part of some of the sockets. If the clay also at the upper part of the sockets dries, it becomes porous and shrinks, or tends to shrink, to the extent of $\frac{1}{24}$ of its moist dimensions, but this refers to pure clay and not to puddle.

Collar Joint.—This consists in its simplest form of a butt joint encompassed by a collar, as explained under that head. In sewer-pipes the collar must be fixed by being run with asphalte or cement. Such joints are sometimes formed when junctions are inserted, though to prevent the disarrangement of a line one of the pipes is not unfrequently broken out, and a sort of bell-mouthed junction effected with brickwork in cement and a slab or two of stone.

Half Socket Joint occurs in piping when each pipe has a half socket at both ends, by the aid of which complete sockets with two opposite joints are formed as soon as the pipes are united.

Lip Joint.—This is formed between the invert blocks of brick sewers when provided with projecting lips to form bond.

Pipe Joint.—The ordinary connection between stoneware and other similar pipes is a spigot and socket joint, usually termed a socket joint. It is made by inserting the spigot or plain end of one tubular piece into the enlarged end, called the faucet or socket, of another, and almost invariably some kind of stuffing or packing is required to fill up that part of the socket not occupied by the spigot. Socketed stoneware pipes are united with clay, cement, or asphalte, without interference with their fullway, the joint being perfectly well smoothed with a stick, and all protuberances and deposits removed before attaching the next length. The utmost care is requisite in every case to guard against the least subsidence after boning, which can only be obviated by bedding not only the ends or sockets but the barrels of the pipes properly, and packing them well underneath; and not only so, but equal care must be evinced not to injure the pipes with the rammer during the operation of filling in, and not less than six inches of earth should at any time intervene between the pipe and rammer head. Cast iron pipes must be leaded and surrounded with concrete.

Socket Joint is described in the previous paragraph.

Stanford Joint is a socket joint between two earthenware pipes made to fit exactly by running into the socket, and round the spigot after the pipes have been burned, rings that are exact counterparts of each other, and composed of a material consisting of ground earthenware pipes, sulphur, and tar, which immediately sets hard and adheres firmly to the pipe. No cementing medium is required, but a water-tight and air-tight joint allowing considerable motion results from merely inserting one end into the other, and painting or not, as thought desirable, with tar or grease. The pipes fit mechanically together, but care must be taken that they are laid on a firm bed, or one that will not become hereafter treacherous, otherwise a settlement of a portion of the sewer might cause the joints to draw. Messrs. Doulton & Co., of Lambeth, are the makers of the pipes.

Tight Joint is one that is impervious to air or fluid under any pressure likely to be applied to it. The socket joints of stoneware drain or soil-pipes out of the ground and inside dwellings ought to be stopped with some material such as asphalte, that is not porous nor likely to become so from continued exposure. It is very difficult to make a joint strictly tight, owing to the imperfection of labour, the alterations in the form of bodies brought about by external forces, shrinkage, and the influence of the weather, to which must be added other causes arising from porosity, from the difficulty of securing in many instances perfect adhesion, and from the subtlety of the wind in preventing a vacuum. The thorough ventilation of drains and soil-pipes, which latter, by the way, ought always to be fixed outside buildings, is about the best means of indirectly conferring upon them the incalculable advantage of tight joints.

SECTION II.

BRICKLAYERS' JOINTS.

Angular Joint is one in which an angle is formed in one of the bricks, as exemplified in the cutting called a birdsmouth, Fig. 1.

Arch Joint.—This usually radiates. In the so-called French or Dutch arch, however, which is only to be tolerated when built in cement, the joints run parallel to the skewbacks. Straight arches have nearly flat soffits and quite flat backs, and their joints radiate to the apex of an equilateral triangle of which the soffit forms the base. They are safe with a wrought chimney bar with bent ends below them for support. A gauged arch is one formed with cut and rubbed bricks set in putty, by which very fine joints are obtained not exceeding $\frac{1}{8}$ inch in thickness, and all accurately radiating to a centre. This kind has a slight camber amounting sometimes to more than an inch per foot of span. The joints of relieving and other rough arches, as well as all those which are turned with ordinary bricks, are wider at the back than at the soffit and are usually turned in half-brick rings, but where through the curves being quick the joints are much wider at the extrados they can be packed with pieces of tile or slate. When turning arches in half-brick rings the first ring requires thin joints, the next thicker, and so on, in order to prevent the rings separating at the collar joints when settling, and where the joints of the rings coincide headers may be built in to form a bond between the rings.

Asphalte Joint is a contrivance for keeping out damp, and is made by applying a layer of hot tough asphalte from $\frac{3}{8}$ inch to $\frac{1}{2}$ inch thick along the top of a course. A good plan in important works is to only half flush the joints at the damp course with mortar, then pour in the asphalte and lay the next course of bricks hot, after which half flush with asphalte. If the walls of a building are not covered by the roof, a damp course is often as much needed at the top as at the bottom. Asphalted felt is used as a seating or bearing for wood or ironwork on stone templates, and is found advantageous in checking vibratory motion produced by percussion.

Back Joint is a joint parallel to the face of the work or any inside vertical joint that is not a cross or transverse one. In footings it should be always kept as far from the face as possible.

Bastard Tuck Pointed Joint.—This is one kind of pat or tuck-and-pat joint, and results from a description of pointing, which differs considerably, however, from common tuck pointing. Whilst the latter consists of very narrow sham white joints of pure white lime putty projecting from a wall face which has been brought by rubbing and colouring to one uniform colour, bastard tuck pointing is produced by forming on the ordinary stopping a projecting ridge about $\frac{3}{8}$ inch wide, having clean cut and parallel edges, and which is not pure white like the other, but of the same colour as those parts of the joints that do not project.

Bead Joint occurs occasionally in brickwork, and is formed by finishing off the pointing stuff with a projecting convex surface in lieu of a square-edged ridge.

Bed or Bedding Joint.—This lies between the courses, or between the bottom course and the bearing stratum, and is that on which almost every part of a construction rests. It is exposed to perpendicular pressure, for the better distribution of which plates or templates are built in to afford spreading bearing surfaces to carry girders, joists, &c. Sleepers, lintels, plates, and templates, wood bricks, and bond timber are generally bedded in mortar, window and door frames in lime and hair or cement, and wood slips dry. Most of these depend to a great extent upon the quality of their bed joints, both for their own firmness and for the degree of shakiness they impart to a building. In foundations where parts start from different levels, those brought up from the lowest should have close joints, or be built in cement or hard-setting mortar to secure an equality of settlement, and in all cases footings should be carefully bedded so as to afford a solid and unyielding base. Iron columns, girders, or pilasters built into brickwork require sound York, or other hard bedstones or templates of a thickness equal to two or more courses, and of as many feet super as convenient to get a wide bearing the full width of the wall. These must be tooled true and dressed fair on the top to take the ironwork, and solidly bedded with full square beds perfectly level in good mortar or cement; and if any perforation, or sinking, or joggling is required for snug, or corking, or flange on column, or rivet on girder, or if any dowel is to be leaded into the template to fit into a slot in the girder, it should be done before bedding the stonework. Girders carrying brickwork are sometimes covered with a course of hard flagstones bedded in cement of the width of the intended wall. In the same way the caps of columns and pilasters must be truly fitted in a workmanlike manner to overhead stone blocks or landings dressed fair and mortised for corkings. Heavy timbers are secured to brickwork by building in stones pierced with countersunk holes for the reception of bolts with heads downwards every few feet, the shanks being built in as the work proceeds, and the timber bolted thereto with countersunk nuts and washers at the proper level. Ironwork bedded or inserted in brickwork does not injure it. In all cases when new work is bedded on old, the latter must be brushed and well wetted before spreading the mortar.

Bevel or Bevelled Joint is one in which the plane of the joint is oblique to the face of the work.

Birdsmouth is any re-entering angle notched out of a brick by way of a moulded ornament, or it is a similar notch required for an inside joint; as, for example, when a wall does not return square in the case of a skew or squint quoin, as in Fig. 1.

Fig. 1.

Black or Blue Pointed Joint is made by using for pointing stuff smith's ashes or coal dust instead of sand.

Bond or Bonding Joint.—This is tantamount to a broken joint, since bricks to form bond must be laid with systematic lapping, so that no two vertical joints in adjacent courses can appear in the same straight line, because they do not in fact lie in the same plane. In the same course, however, the cross or transverse joints need not be broken, for work to be well bonded only requires the vertical joints to be well covered by the superimposed bricks. Owing to straight cross joints, however, not always being considered sufficiently weatherproof, some break them, and thus abstract from the work a certain amount of longitudinal tie. As the only condition exacted by bond is the discontinuance of the vertical joints of each course at the rising one, there are, as might be expected, several ways of arranging bricks by which this requirement can be more or less perfectly fulfilled. Amongst many varieties of bond made use of the following may be noticed as the most generally adopted.

Angle bond occurs at skew quoins or oblique angles requiring cut bricks and offering great temptations to hide very rough cutting with flushing. It may be here observed, that in order not to leave voids, closers of some shape or other are required to get the lap in all bonds. They should never be set at a less distance than half a brick from any opening or quoin. Chimney bond consists of nothing but stretchers, and is therefore only half a brick in thickness. Diagonal bond is useful for maintaining longitudinal strength in a thick wall, and consists of ordinary face work with a hearting laid raking or diagonally at an angle sufficient to admit the bricks without snapping, as shown in Fig. 2. English, or as it is often called old English bond, and sometimes King's bond, presents on each face a course of headers alternating with a course of stretchers.

Fig. 2.

Here, obviously, it is practicable to vary the arrangement by having two, three, or more courses of headers to one of stretchers, still maintaining an excellent bond. Flemish bond shows on each face a series of courses each formed with alternate headers and stretchers. Garden wall, or running bond, Fig. 3, consists of one header to three stretchers on each face and every course, the headers of each course being in the centre of the stretchers of the adjacent courses. Heading bond is made up

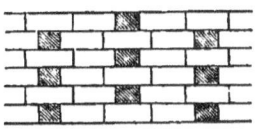

Fig. 3.

of courses of headers only. Herring-bone bond resembles diagonal bond excepting that the hearting is laid in the direction of both diagonals. To these may be added cement bond noticed under Cement Joint, and hoop iron bond, which does not consist of any particular arrangement of bricks, though its province is to obtain longitudinal strength, which is the boast of old English. This bond is formed by building in about every twelfth course one row at least of 1¼ in. hoop iron tarred and sanded to each half-brick of thickness. A 14 in. wall would therefore have three parallel rows. The hoop iron must be properly bent, lapped, and hooked at junctions, and wherever necessary to maintain the tie. It is often bedded in cement. Tyerman's patent variety is much used, having a good grip through each edge being notched at intervals into vertical teeth in opposite directions. When brickwork is very thick the faces are laid in the approved bond and the interior is filled in by larrying, which consists in spreading a thick bed of mortar and pushing each brick along it into position, by which means all become snugly bedded and their sides well surrounded with mortar. The cross bond of hollow walls is necessarily imperfect, the 2¾ in. or so hollow space being crossed by iron ties either wrought, galvanised, or cast, built in every fourth course excepting near the openings, quoins, and angles, where they should be placed every alternate course. Their usual horizontal distance apart is about 18 in., and no mortar or rubbish must be allowed to rest upon them.

Broken Joint.—This is explained under Bonding Joint, and may be made by lapping or crossing. It is essential in all good bond, and is introduced wherever possible at all the parts of the composite carcase, whether of stone, brick, wood, or iron.

Cement Joint is generally made with Portland cement and sand quite free from impurities where strength is required. For any other reason but that of economy the less sand mixed with it the better. Work hurried forward should be executed with quick setting cement, otherwise there will be unequal settlement. Neat Roman cement sets in about fifteen minutes, but gauged with an equal quantity of sand requires much more time. For a rod of reduced brickwork, about 75 bushels of cement and sand are required—and about 150 gallons of water and 37½ bushels of cement, supposing the Portland cement and sand to be in equal proportions. Cement bond is nothing more than three or four courses of brickwork in cement, which should be kept back from the face so as not to interfere with the pointing. When lines of hoop iron are built in it becomes hoop iron bond.

Chimney Pot Joint is formed by setting the pot or can, when of the red earthen or other similar kind devoid of collar, on the top of the orifice of the flue in cement mortar and well flanching it therewith, and sometimes additionally protecting the joint from the ingress of water by means of a flanching or covering of tiles bedded in cement. Terra cotta and other moulded or ornamental pots are often provided with a projecting collar near the base, so that in their case the base is stepped into the orifice and the joint made tight by flanching in the usual way.

Close Joint.—This does not exceed in thickness ¼ in., and always requires good workmanship. Brick arches should be close jointed at the intrados. In this trade close joint is perhaps more correctly termed thin, whilst fine or very fine more particularly applies to those common to gauged work, which ought not to be more, and are sometimes less, than ⅛ in. thick.

Coarse Joint.—This may be defined to be one thicker than is agreeable to the eye. The outside limit of thickness in good brickwork may be set down at ¼ in., whereas $\frac{6}{16}$ in. may be assumed to be a fair working average for ordinary facing. A thickness of ½ inch and upwards may therefore be reckoned coarse, though frequently met with, especially where deformed and irregularly shaped bricks have to be worked in. It is necessary to remember, however, that where strength and not appearance is the desideratum, if the mortar is stronger than the bricks there can hardly be too much of it, whereas if it be weaker there can scarcely be too little. On the other hand, where appearance only has to be considered thick joints are inadmissible; and so they are, having regard to the quality of mortar customarily mixed up and used by speculative builders, for it is not stiff enough nor adapted to withstand the squeezing produced by the weight of several courses while yet in the green state. Thick joints induce unequal settlement, are more pervious to rain, and require repointing oftener than thin ones. Sometimes when the old standard of four courses to a foot or thereabouts has been insisted on, the bricks not coming up to the expected thickness have caused the bricklayers to make coarser joints than the designer of the fabric desired. When there is any doubt as to the thickness of the bricks, it is better to stipulate that no four courses shall rise more than a definite extent besides the height of the bricks, and to avoid thick joints the bricks must be equal sized and well shaped.

Collar Joint in arch work is a continuous bed joint surrounding a ring course. It ought to be of regular and equal thickness between each course, otherwise the thickest will shrink most and perhaps produce an unsightly crack, the sure indication of weakness. In some descriptions of work, especially if underground and exposed to an influx of water, it is politic to execute it in Portland cement with a thickness of an inch or so, in which case the same cementing medium will be required throughout, for cement has a tendency to swell instead of shrinking like mortar.

Coursing Joint is the joint between two courses. In arch work it lies between two string courses.

Cross Joint is one which stands at right angles to a coursing joint. Perpends are formed by a series or line of cross joints truly plumb or perpendicular to the horizontal bed joints. The inside upright or vertical joints are either longitudinal or transverse (if not raking), the former being styled back and the latter cross joints, and both of these are in reality butt joints. In good work all should be flushed solid. In a cut and gauged arch the joints parallel to the soffit or intrados are cross joints. It may be added that bricks are said to be laid cross joint when all the vertical joints of a course break joint with those of the next.

Cut and Struck Joint is described under Struck Joint.

Diagonal Joint occurs in raking bond, and in single and double herringbone paving, in diagonal quarry paving, &c.

Dipped Joint is made in gauged or rubbed work when setting the bricks in lime putty reduced to the consistency of cream. The bricklayer dips the side of the brick which is to become the bed into the putty so as to allow only just enough to adhere to it to make the joint. It is then bedded and driven close home.

Dovetail Joint.—Bricks cut and arranged as in Figs. 4 and 5, form what are called dovetailings, the dovetail joints enclosing them being readily recognisable.

Fig. 4.

Fig. 5.

Drawn Joint is one that is filled in flush and smoothed with the trowel and drawn with the jointer and jointing rule. Before the mortar sets the rule is pressed firmly against the work, or a line may be stretched across it instead, and the jointer is run along to mark the centre and equalise the thickness. In whatever way the joint may be finished the distinguishing feature of a drawn joint is the use of the jointer.

Face Joint.—This shows on the face of work and may be neatly cut and struck, or left flush, or tuck pointed, or finished in any of the other appropriate ways noticed in this section under their specific heads.

False Joint is a face joint occurring in tuck pointing or in the horizontal joints of gauged arches, or in any other way to imitate a real joint, which under these circumstances must be concealed by daubing if too close to the false one.

Filleted Joint is made when fillets of bricks are built up with the substance of the brickwork to cover the edge of the slating instead of depending upon cement or mortar fillets where flashings are not used. The brick fillet oversails about $1\frac{1}{4}$ inch, just above the slating, the joint between the fillet and slating being made good with cement.

Fine Joint is a very thin joint only found in cut and rubbed work when ordinary bricks are used, and often goes by the name of Dipped Joint.

Flat Joint is formed by pressing the mortar flat as in a flush joint, for which it is another name, and marking the courses with the edge of the trowel. It is common in pointed work and is made in three different ways. The pointing stuff is either laid on with the trowel and cut off at the top only, or it may be cut off both at the top and bottom, or else it is filled in flush and a line run along the centre with a jointer, in which case it becomes a drawn joint.

Flat Joint Jointed.—This has been described under its alternative name of Drawn Joint.

Flat Struck and Jointed Joint is the same as a Flat-ruled Joint, and is executed in some localities by drawing the joint up full without cutting the top off and afterwards running the jointer along the middle with a straight edge. It is advisable to strike joints along their upper edge so as to make the sloping surface a sort of weathering, but the reverse is the common practice.

Flush Joint is one in which the mortar is in the same plane as the face of the bricks. It is made by pressing the mortar flat on the joint, and sometimes in addition the point of the trowel is drawn along the top and bottom, as in Fig. 6.

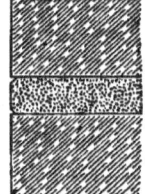

Fig. 6.

Flushed Joint. — Coursing joints are flushed solid when the mortar in the inside joints is brought up flush to a level surface throughout. When bricks, however, are only buttered instead of being laid on a bed of mortar properly spread, flushing will not suffice, but grouting must be poured in to reach the vacuities. In some cases thin mortar is used for flushing so as to dispense with grouting.

Flush Jointed Joint is executed by drawing the point of the trowel along the top and bottom on laying the bricks, and afterwards when the mortar is sufficiently set the middle of the joint is flushed flat by running the jointer along a straight edge. Or it may be made by drawing the joint up full without cutting the top off and similarly using the jointer, as shown in Fig. 7.

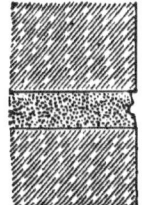

Fig. 7.

Full Joint occurs when the mortar approaches the face of the work sufficiently near to be flush. If it does not it is either too full or not full enough, which necessitates raking out the excess or filling in the defect with stopping.

Grouted Joint.—This is explained in the Mason's Section and is employed in order to fill the open joints and splits in the interior of brick walls which flushing will not reach. The mortar should if possible be quick setting. As Portland cement has the power of rejecting a superfluity of water its use may be countenanced where saturating the brickwork is of no consequence; and in considering the value of the practice it is obvious that it ought never to be regarded as anything beyond a bad alternative for wet bricks and stiff mortar, the odds being in favour of grouting when dry bricks are preferred. The utmost care, however, is needed in grouting the oversailing courses, &c., when the overhang is considerable, owing to the chance of the lime swelling, or of its being drowned, or of the bedding being loosened. In paving with bricks, tiles, tesseræ, quarries, &c., these are first laid on a bed of cement, screeded or floated to the approved fall, and then the joints are run with cement grout, all excess being completely wiped off before setting, else it will be very difficult to remove. Stables are paved with Musgrave's adamantine clinkers placed on edge with open joints upon a bed of concrete, laid with a slight fall in every direction to the surface drain, and also in such a way that the surface of the finished paving may be a little above that of the drain. The open joints are then run with cement grouting, the surplusage being at once cleaned off.

Hacked-out Joint occurs when covering the exterior of old buildings with stucco, the joints having to be well hacked out and wetted to give the stuff sufficient hold or key.

Heading Joint is a term only used in arch work by the bricklayer. It is exactly the same as the corresponding mason's joint, occurring between the ends of the bricks of the ring courses.

High Joint is another name for flush jointed joint.

Hollow Joint occurs when the inside of work is not flushed solid, and is usual in fence walls. Bed joints when not subsequently filled up with grouting are also left hollow when different thicknesses of bricks have to be worked in together, because the bricklayer cannot spread out his mortar to form a bed for the thicker bricks, else he would lose his gauge. Consequently he draws a little mortar along the front and back edges and lays the bricks with a hollow joint. This kind of joint, however, is necessary in some situations, as for instance under window sills, which should be set so as to bear on their ends, with a hollow joint between them to allow for the piers settling more than the work under the apertures.

Jointed or Jointer Joint is formed by drawing the tool called a jointer, which is often nothing more than a suitable piece of old iron, along the joint so as to strike a central indented line as shown in Fig. 8. The similarity between a drawn and a jointed joint is here apparent.

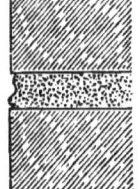

Fig. 8.

Keyed Joint is a modification of a flush joint produced by grooving its centre with a jointer, or key, or pointer having a rounded edge, which is rubbed along the joint so as to press the mortar back and leave a slight concavity.

Lap Joint.—As already pointed out, this is the fundamental joint of all proper bond. In a slate damp course all joints must lap to retard water rising by capillarity, and when the walls are finished off with a parapet or coping, an upper damp course is advisable to prevent water soaking down.

Lip Joint.—This occurs between blocks of terra cotta when used for facing or for sewer inverts, &c., and is described under the name of that material.

Loose Joint is easily imagined without explanation. It is characteristic of careless beam filling, indifferent bricknogging, ill-laid paving, &c.

Mitre Joint.—This is more fully explained under the Mason. It is formed in brickwork in panelled walls, chimneys, &c., as shown in Figs. 9 and 10.

Mortar Joint is at once the commonest and most mysterious of all joints; most mysterious because neither the reasons of its setting nor the time required for the purpose can be even now predicated with certainty, and because there is no satisfactory agreement amongst the leading

Fig. 9. Fig. 10.

architectural and engineering experts as to the best and simplest way of making it properly, notwithstanding the thousands of hands daily employed in its fabrication; and, finally, it is most mysterious because in some unaccountable way rubbish heaps rise up and disappear in the vicinity of the spots where the mortar men are at work. A mortar joint, however, is commonly formed by laying bricks with a composition of lime and sand in the proportion of 1 part of unslaked lime to 2 parts of dry sand (or more or less as preferred) mixed with sufficient clean water to slake the lime and work and temper the mass to a stiff or sloppy paste, as thought most desirable. When more than 2 parts of sand are used it is questionable whether under the usual conditions of mixing it becomes properly incorporated or blended with the lime. The mixing should be done upon boards or a brick or stone platform where no dirt or gravel can get worked in with it. Three cubic yards of mortar are allowed to a rod of reduced brickwork, in which about 192 gallons of water are taken up when stone lime is used. The lime ought always to be more or less hydraulic, else it will not set nor be exempt from injury by frost. It should be fresh burnt, free from core or lumps, and cautiously kept from exposure until used. The sand requires to be clean and sharp. Blue lias lime may be well used hot from the kiln; it must be thoroughly slaked and mixed in small quantities as required with just sufficient water to bring it to a proper consistency, and on no account must it be allowed to partially set before using. 24 hours after being made, this sort of mortar, if the lime be technically of poor quality, will probably set in air, and if left undisturbed water will have no effect upon it. The thickness of a mortar joint in ordinary brickwork should not exceed $\frac{6}{16}$ in., though excellent work is known to exist with much thicker joints. The less that is used of inferior mortar the better, provided the bricks are pretty square and wetted till soaked, and that the mortar bed is thick enough to prevent the bricks touching, which necessitates a minimum theoretical thickness of $\frac{1}{8}$ in. or a practical one of $\frac{1}{4}$ in. A joint of this thickness has the advantage also of looking well, especially if throughout the work they all maintain the same size.

Parallel Joint.—One that is contained between parallel planes. To secure this in arch work the bricks must be wedge-shaped.

Pat Joint is an abbreviation for tuck and pat joint, and is the result of tuck or bastard tuck pointing.

Perpendicular Joint is the same as a vertical, or cross, or back joint, or one that is at right angles to a coursing or horizontal joint. In face work, "keeping the perpends" signifies that the visible part of the perpendicular joints are kept exactly over one another, or in the same vertical line throughout the whole height of wall, without which precaution the courses would overrun and bats would have to be worked in, which would greatly detract from the look of the work and the efficiency of the bond.

Piecing Joint occurs when new work is built up against and bonded with old, which must be toothed or racked back, and well brushed and wetted. A chase can be cut forming a slip joint where the new work meets the old at an angle, but in no case must the new by its sinking be allowed to hang upon or strain the old. Iron ties may be built in as the new work is carried up, but not fixed to the old until the former has settled down to its bearing.

Pointed Joint.—This is described generally in the Masons' Section, but various methods of executing the joint will be found under their specific names in this list, as, for instance, under that of bastard tuck pointed. Pointing to old work includes stopping, which consists in refilling the joints after raking out and cleaning. In all cases the work should be well broomed and wetted before inserting the pointing stuff.

Putty Joint.—Gauged work is set in putty, made by dissolving lime in water, running it through a sieve, and bringing it to a thick creamy consistency; the joint also finished off by tuck pointing is indicated if not made with a composition called putty, and which is either white, black, or red; it is laid on after the work is stopped. The white putty is a mixture of chalk lime, which must be well screened, and silver sand or marble dust, a small quantity of oil or size being added to the lime whilst dry. Black and red putty are both made with grey lime and fine well-washed sand, the former being coloured with lamp or vegetable black, and the latter with Spanish brown. Vegetable black mixes the most readily with the other ingredients.

Radiating Joint occurs in arch work, and in the facing of thick brick piers, &c.

Raised Joint is the same as the joint of similar name made by the mason.

Raked Joint is met with when the joints are raked or hacked out with a hooked iron point, previous to outside plastering or inside rendering where they have not been left sufficiently rough to afford a key; the same term is also applied to joints in face work which are slightly sunk or recessed for effect or to give light and shade.

Before pointing old work, and previous to tuck pointing, &c., it is necessary to rake out and clean and brush the joints to a depth of ¾ in., or thereabouts.

Recessed Joint is one raked back, or grooved for effect. One variety has been described under Keyed Joint. Though often admired for the play of light and shade which it affords, it is on the whole objectionable, owing to its offering in the aggregate a large surface for the lodgment of rain.

Rough Joint occurs when the joints are left unstruck, and protruding to afford a key for the rendering.

Rubbed Joint.—This is formed between bricks brought down by rubbing on the rubbing stone to any required bevel. Bricks have frequently to be cut before they are rubbed.

Saddle Joint occurs between a saddle piece and the overlying brickwork, and is similar to the corresponding joint described under the mason.

Shift Joint is a broken joint, or one that is not continued straight, but given a shift to one side to break joint.

Slip Joint is formed when a chase is cut in old work for piecing new work to it, in order that a straight joint may not show, and that the new work may settle without hanging upon or straining the old.

Spring Joint is a loose joint.

Straight Joint.—A long joint proceeding in the same line or plane without a break. When running in a slanting or vertical direction it is a source of weakness, and almost any artifice is allowable to curtail or avoid it if there be no orthodox way of escaping it.

Struck Joint is formed by drawing and pressing the point and edge of the trowel along the mortar joint until it is quite smooth, and forming a slightly sloping surface by keeping the trowel well in towards the top edge of the joint. The bricks ought to be laid perfectly level on bed, and in a true plane as to face, with the mortar left as full on the face as possible to allow for pressing in. As soon as the course is laid out all holes in the joint must be stopped, and then it is ready for striking, which is done by smoothing the mortar as above described. Very often the trowel is run along the bottom edge instead of the top, but whichever plan is adopted the joint ought to be as full as possible without projecting. When it slopes outwards and downwards by striking the joint according to the first method, there is little chance of any wet lodging; and if the remaining or most projecting edge, in either case, is cut off straight with the trowel, either with or without a rule, the joint is sometimes said to be struck and cut, though the term "struck" includes cutting.

Sunk Joint is the same as a recessed joint, or one drawn with a thick jointer.

Terra Cotta Joint is one peculiar to this material when moulded in alternate thin and thick blocks for facing, which are usually about 1½ in. and 6 in. thick respectively. Both sizes have corresponding but reversed lips, by which means the thicker blocks hold the others in place as in Fig. 11. The plain butt, however, is as often as not used, and blocks for columns, and other features have joggled or dowelled or grooved and tongued joints.

Fig. 11.

Thick Joint.—The remarks under Coarse Joint are here applicable, for coarse and thick joints are interchangeable terms. What would be considered thick in facing, however, are not so in hearting or backing, and what were once thin in the former situation soon become thick through weak mortar and repeated pointing. The safe crushing strength of good mortar is certainly not less than 1 cwt. per square inch, and provided such were used, and the work built solid, and not carried up too fast (say not more than 2 or 3 feet per diem), properly distributed pressure would doubtless be obtained by combining thin facing with thick inside joints when laid as described in the succeeding paragraph. At all events, whether they are contrary to sound construction or not, having regard of course to the mortar of the present day, a goodly number of thick joints lie concealed in every structure.

Thin Joint is one ranging from $\frac{1}{8}$ in. to $\frac{1}{4}$ in. thick, the minimum depending only upon there being no "brick and brick," but sufficient thickness to prevent touch, which is assumed in practice to be about $\frac{1}{8}$ in. with the very best bricks, and $\frac{1}{4}$ in. with picked stocks uniform in size and regular in shape. There must always be a certain margin left for the imperfection of workmanship on the one hand, and for slight distortion of the bricks on the other. The thinner the joint the more necessary is it that the bricks should be soaked before laying, so that the moisture essential for setting may not be abstracted from the mortar. Less settlement occurs with thin than with thick joints, for the shrinkage of common mortar is proportional to its bulk, but in good work, where facing bricks are used, the difference is neutralised by the bricklayer knocking the bricks in the backing harder down than the others, whilst every course is flushed full and no snapped leaders allowed. In massive engineering structures, it is feasible to get $\frac{1}{4}$ in. bed joints throughout, with sound hard burnt stocks of good shape. The bricks, however, as already noticed, must be thoroughly soaked, and if not grouted, flushed solid with sufficiently thin mortar to penetrate into all vacuities. The action of the weather is less prejudicial to thin joints. Kiln bricks are usually made larger than clamp bricks or stocks to obtain thin joints in face work, since four courses of rough stocks, owing to their irregularity of shape, sometimes rise from $1\frac{1}{2}$ in. to 2 in. more than the height of the bricks. Owing also to the unequal thickness of the commoner kinds of bricks, joints have sometimes to be made here and there very much thinner than intended, otherwise there would be no means of keeping the thicker bricks down to the required gauge. Bricks with frogs or kicks can usually be laid with thinner joints than those without them, but to insure filling the frogs they must be laid uppermost.

Tie Joint is a bed joint strengthened by means of hoop-iron built therein for a tie. It has already been described under bond. In the case of chimneys rising above the roof, crossings of the same should be laid along the withs and lapped to the tie in the chimney walls. It may be inserted every fourth course, or oftener if thought desirable.

Toothed Joint is formed when piecing into toothings either made or left for the purpose.

Transverse Joint is the same as a cross joint. Its plane is usually perpendicular to the face of the work, and it shows on the face as a vertical joint perpendicular to the bed joint.

Tuck Pointed Joint.—This is a false or sham joint, occurring principally in brickwork. There are different kinds, but that commonly made with ordinary stocks is done as follows. First of all, the real joints are raked out and the whole of the face is cleaned and rubbed down with a soft piece of the same kind of brick, the dust is brushed off, the wall wetted, and the colouring to bring the light and dark bricks to the same hue, and which is a weak solution of green copperas, is applied. A length of 7 or 8 feet by 5 feet deep is then stopped in with a stopping composed of one part of lime putty and three parts of fine washed sand, coloured with yellow ochre to match *when dry* the colour of the bricks. In other cases the whole front is stopped before pointing is begun. The part thus prepared is then rubbed down with a piece of dry canvas to bring joints and bricks to the same colour, and narrow grooves accurately gauged (if thought necessary by means of a gauge rod) are run with a long straight edge where the false joints are to appear. A narrow strip of white lime putty, composed of chalk lime and a little silver sand, is then laid perfectly level along the grooves with the rule adjusted to the gauge marks, the rough edges being cut off with the frenchman, thus leaving a white narrow projecting ridge neatly pared to a regular width of about $\frac{3}{16}$ in. The vertical joints are similarly done, and kept perfectly plumb all the way down, so that the work when finished appears to be well built with picked regular-shaped equal-sized bricks with very fine joints.

Upright Joint is either a cross or a back joint.

Vertical Joint.—This includes transverse and back joints, and is obvious from its name. In depositing concrete it is advisable to bevel off the end or edges well when knocking off work, so that there may be no vertical joints between the new and old deposits. To insure a monolith, moreover, the concrete should be spread and rammed in layers not exceeding 1 foot in thickness, the upper surface being well watered if dry when the next layer is about to be tipped in. In laying bricks it is customary merely to plaster a little mortar along the edges of the bricks which show outside, but by subsequently filling up the vertical joints of each course by the operation known as flushing, all the inside cavities disappear, and the work should become solid throughout. If the mortar, however, is not properly spread over the bed, grouting will be required as well, which many consider prejudicial to good work as interfering with the proper setting and adherence of the mortar. Flushing with thin mortar is a more scientific expedient.

Wall Joint.—Another name for a back joint. Its thickness is usually about $\frac{3}{8}$ in.

Weather Joint is one designed or protected with a contrivance to keep out wet, or wind and wet. To this category all the external joints of a building belong, or rather should belong, for it will obviously depend upon the degree or quality of workmanship displayed in their formation, whether they will stand or give after the trying ordeal of a few rounds of the opposite extremes of stifly hot, and boisterously cold weather. As a general rule, bevelling or slanting off a surface downwards and away from the joint more or less weathers it, owing to water finding its own level. Mortar joints therefore should be struck with a surface sloping inwards towards the underside of the course above to allow the rain to glance off. For the same reason, chimney pots and roofing are flanched, and walls coped, &c. In hollow walls a piece of 5 lb. sheet lead bent to form a sort of small gutter, and extending about 4 inches beyond each side, is built in over all door and window frames to prevent any wet reaching them that may find its way through the outer thickness.

White Joint.—One formed with ordinary mortar as distinguished from blue mortar. Or it is made by pointing with white putty, which is a stiff paste composed of the purest chalk lime, and silver sand, or else marble dust. The latter ingredient is the best of the two if reduced to powder by heat and well screened. Where cement bond is used it is necessary to rake out the cement before setting, or to keep it back from the face in order that the joints may be pointed similarly to the rest, which would be most likely executed in white mortar, though a good deal of blue or black is used.

Wood Joint occurs when wood occupies the place of the usual mortar joint between courses of brickwork, in order that woodwork may be fixed thereto. It is usually formed by laying or building in the wall a pallet or wood slip the length and width of a brick and about $\frac{3}{4}$ in. thick.

Such slips are now preferred to wood bricks, as they shrink less and afford better hold by not becoming so loose. Ranging bond, which often consists of parallel slips of the same width and thickness built in dry the whole inside length of the wall at distances apart varying from 1½ ft. to 3 ft. to receive match boarding, battening, lining, wainscoting, &c., constitutes also another kind of wood joint.

SECTION III.

MASONS' JOINTS.

Angular Joint occurs in joggling, back rebating, &c.

Arch Joint is a bed joint between two voussoirs or arch stones. It usually radiates, and is always more inclined to the horizon near the key than at the springing, and hence is more compressed as it nears the latter position. The joints at the crown only begin to feel compression when the key is driven home, but are not actually compressed until the centre, which supports the whole weight of the crown near the key, is eased or struck. Unless the centre, therefore, is rigidly and properly braced, and the arch joints dressed perfectly straight and true, and made full and close, and the stones brought well home to their beds with a maul, the easing of the centre will cause the arch ring to alter its form and the joints to open at the soffit near the crown and at the back at some part of the haunches. Under these circumstances before striking the centre all open joints should be well wedged with pieces of slate and grouting. Plates of lead have been frequently introduced into arch joints to obtain an equal distribution of pressure and prevent the opening of the joints, but not always with the same success. Some allusion to the practice is made under Lead Joint. Stone and metal plugs or joggles are much used at the joints of ordinary arches, and to prevent sliding the beds used to be sometimes embossed and hollowed. All arch stones should, if possible, be either headers or stretchers, and the arch turned in one ring of equal-sized blocks, with the exception of the faces, which to preserve the bond require stones of two different lengths. Gothic arches often have a central joint which does not radiate. In a skew arch the joints usually form a spiral surface of uniform twist that radiates from the axis of the cylinder coinciding with the soffit of the arch.

Asphalte Joint occurs in masonry as a damp course, as described in the Bricklayers' Section. Iron cramps employed to bind together two blocks of stone put in hot and run with asphalte are secure against oxidation.

Axed Joint.—This results in worked granite when the surfaces of contact are first broached with the pick, and then finished for a certain depth with the axe or heavy hammer with sharp pointed edges required for bringing them to this sort of finish. Further particulars will be found under Fair Axed.

Back Joint.—Generally the junction of the back of a stone with what it butts against, whether wood, iron, brickwork, or another stone. It is a joint that often needs tooling or dressing, and it may be found at the junction of stone steps, either with or without a rebate, or at the inner surfaces of chimney-piece jambs when jointed to return pieces, or at the meeting seam of a stone step with a boarded floor, &c. It always occurs behind the face of the work, excepting in an arch where it runs parallel or thereabouts to the soffit and forms the bed joint of the next ring or spandril. In ashlar facing the back joints are usually roughly drafted square with the beds.

Back Rebate Joint occurs in masonry when a back rebate or birdsmouth is formed, either to receive the sally of another stone, or to conceal a junction, as in the case of stone door and window architraves or dressings, which by such means are made to overlap and conceal the unsightly straight joint that would otherwise be conspicuous.

Bag Joint.—When bags of cement are sunk to the bottom of cylinders and rammed down to form foundations where loose cement would be washed away, they virtually become in time solid blocks separated by the bags only. Bags of concrete used likewise for similar constructions may be readily packed together and well jointed by divers so as to make solid work. In both these instances the junctions between the indurated blocks may be appropriately termed from analogy bag joints.

Bed or Bedding Joint.—Much that has been said under the Bricklayer with regard to this joint is applicable here also. In order to withstand perpendicular pressure the joint should be left full and square for the whole depth, and whenever ironwork is built in the strictest attention has to be paid to verticality of axis and horizontality of bearing surface. In ashlar a tooled or plain draft is worked round the edge of the bed out of winding (except in the case of skew arches, &c.), the rest of the bed between the drafts being only roughly, though neatly, pointed or picked to afford a good key, care being taken that nothing protrudes above the drafts. Usually the bed is boasted, and often accurately worked so that no part lies more than $\frac{1}{8}$ in. below the edge of a banker rule applied to opposite drafts. The block should be set firmly without pinning or levelling upon the cementing medium laid over the upper bed similarly wrought of the course below. Good work is executed with lias lime and Portland cement grouting. Foundation stones and base stones consisting of flag stones or landings running under piers require level beds and surfaces, and fair joints. Bed stones, base stones, and templates have been noticed under Bricklayer. Post stones for heavy iron columns should be thick, and have a base of about six feet super. The top beds of all stones forming a bearing surface for setting or fixing the longitudinal or vertical iron or timber elements of a structure must be dressed fair, and the bottom ones left square and full. Drums, or frusta, or tambours of columns unless bedded accurately, will flush or fracture at the arris. A sheet of lead, cut rather less than the bed so as to leave about an inch between it and the outer edges of the stone, prevents flushing or the splintering off of spalls. When Portland cement is used the joints should be raked clean out all round about an inch in depth before it sets. One mode of applying cement grout is as follows:—Four small pieces of sheet lead about 3 in. square being disposed between two adjacent frusta so as to keep the beds asunder and equally bear the weight, the joint is pointed with mortar, leaving a couple of small holes. A half-inch hole is then drilled from a point about 3 in. above the joint so as to enter the bed 2 in. from the face of the stone. The grout is then to be poured in till the joint is full. The surplus water will escape through the holes left in the pointing, and when the cement is set the joints are to be cleaned out an inch from the face, and stopped and pointed as desired. Where granite bearing blocks are carried by iron cylinders filled with concrete or brickwork, such filling should be allowed to project an inch or so above the iron casing, so that the blocks may be wholly bedded thereon without touching the iron. As a general rule, the more important the work the more necessary is it that the top and bottom beds of stone should be carefully dressed to an even surface, so that the layers of fine mortar introduced between them may be of uniform thickness. Other particulars relating to bed joint will be found in the Smiths' Section, as well as under asphalte, bond, cramped, dowelled, lead, &c., in this.

Birdsmouth.—A rebate sunk on the edge of a stone, as well as the joint resulting therefrom, is sometimes so called.

Blown Joint.—A joint may be blown either through the swelling of cement, or in work exposed to the action of waves from hydrostatic pressure on the air confined in the joint.

Bolted Joint.—Massive iron castings intended to act as footing pieces to the ends of arched principals, or as base plates to the ribs of cast iron arches, are held down on masonry foundations by means of holding down or anchor bolts from 2 in. to 3 in. diameter, which are built into the work with heads downwards at points a few courses below the bed joints of the castings, and pass upwards through drilled holes or holes cut in the joints. Similarly, masonry in breakwaters, lighthouses, &c., exposed to the fury of wave action, however well joggled and dowelled, may be reinforced in strength by iron bars or straps connected with rods or bolts passing through drilled holes in the blocks of several courses, and well screwed up with proper nuts. Where concrete is used, rod bolts 3 in. or 4 in. diameter may be conveniently secured to underlying courses of masonry, and left projecting upwards with twisted ends to be irremovably set and fixed in the concrete that is afterwards deposited to consolidate round and about them. Works on a smaller scale require blocks of stone to be mortised for lewis bolts which are run with lead, or else the stones are bored right through to receive larger bolts. In either case the screwed ends of the bolts pass through the bolt holes in the ironwork, and after the bearing surfaces are truly levelled and adjusted as elsewhere explained, the ironwork is finally bedded upon the masonry, and the two are tightly united with nuts and washers. Doorcases likewise are secured to stone door jambs by ⅝ in. bolts and nuts run with lead. Cast iron chairs supporting the feet of iron rafters are bolted down to stone templates, &c., with lewis or dovetail bolts leaded thereto; and amongst numberless other applications of the screw bolt to mason's work may be noticed the securing of guard piles to the face of quay walls, which is well effected by its aid.

Bond or Bonding Joint.—The size of the blocks in ashlar masonry makes the contrivance called bond, described in the Bricklayers' Section, of great value from its contributing to the formation of walls of no mean strength without the adventitious aid of cement, provided only that the stones are set with beds and joints accurately dressed both square and smooth. Bond, through its instrumentality, by enlisting as it were to its service the vertical pressure due to the operation of gravity, causes the weight of each block to fall upon two or more adjacent stones of the course below, by which means they become more or less united; and the same process occurring also with the resultant pressure on each block, not only is the wall rendered firm, but by a precisely similar action the pressure produced by inequality of load is widely distributed before it is transmitted to the ground, so that in fact the strength of ashlar necessarily depends upon the excellence of its bond. The various kinds of masonry, however,—whether coursed or uncoursed ashlar, block-in-course walling, squared rubble coursed or uncoursed, random rubble, or any other description compounded of one or more of these—have but the one common bond of the broken joint. In the present day neither the preparation of the stones nor the way they are dressed with the pick, point, or tool, nor the mode of arranging them in the courses, gives a nomenclature to the bonding as such, but merely to the general disposition of the stones. There are one or two exceptions, however, which are unimportant, inasmuch as the terms are but little used. Cross bond signifies that the cross and back joints break joint at the adjacent courses, as, for instance, when there are two and three stones alternately in the width of each course. Dog's tooth bond is an appellation given to a simpler kind of cross bond by which binders or headers cross alternately from opposite sides and overlap in the centre of the wall for a space equal to $\frac{1}{3}$ its thickness. Notwithstanding what has been said in adulation of bond, it is but fair to observe that no matter how good it is, and however well cemented, it is yet insufficient without other devices, such as dovetailing, joggling, cramping, and bolting both blocks and courses to obtain that approach to monolithic strength which constitutes one of the goals of modern engineering. This is particularly the case when the load or impact does not act vertically. Moreover, good bond, though of the highest importance and even thus assisted, is of little avail without careful bedding, and this again cannot be accomplished without accurately cut beds and joints, fine mortar, and perfectly solid work free from all dry packing, interstices, vacuities, and splits, especially under the through stones, which pass right through from back to front, or other headers sometimes called inbonds or binders, which only lap with their tails to form the cross bond. Unless these rest on consolidated beds, they act with leverage and set up internal dislocating forces. With all descriptions of masonry it is comparatively easy to break the vertical joints in the facing or backing, but the transverse bond being out of sight, and the hearting or filling of thick walls consisting often of rubble, inducements are thus offered to arrange and bed the stones in the interior of walls with an absence of that painstaking and systematic care so frequently observable at the more conspicuous parts. Hence, careful and firm and intelligent supervision is very necessary. The chief feature of good bond in all varieties of masonry for ordinary building purposes, over and above what has been already animadverted upon, is a proper proportion of equal-spaced, equal-sized headers, showing along each course an aggregate length equal to from $\frac{1}{6}$ to $\frac{1}{4}$ of the whole. Where voids occur, the headers should show a combined area holding a somewhat similar proportion to the total area of the façade. This should be supplemented by a fair overlapping of all the stones, and a careful selection of the best and largest blocks for the corners and chief openings. In the case of masonry built to resist percussive action, horizontal bonding courses extending throughout from face to back, vertical cross walls and counterforts are expedients more or less often resorted to, but in all cases the hearting and the contents of the pockets, if intended to add to the strength of the structure, must be either of the best hydraulic concrete carefully deposited and rammed, or else of well-packed and thoroughly grouted rubble. Analogous to the use of hoop-iron bond in brickwork, chain bond is resorted to in masonry. This consists of one or more rows of wrought-iron bars 4 in. by $\frac{3}{4}$ in., or thereabouts, let into grooves in the bed joints and completely embedded in Portland cement. The bars are connected by cross pieces when the walls are thick, the whole being well secured by bolts, or otherwise, at the joints and angles. Chain bond is further noticed under Collar Joint.

Broken Joint.—As will be gathered from the last paragraph, this is the antithesis of a straight joint, and equally necessary in masonry for appearance and strength, excepting, however, at the beds, where custom and convenience connive at long straight unbroken ones. It is obtained by setting each block so that its side, or end, or back does not fall in the same vertical plane as that of the back or joint of a block in the course below, but jets beyond or falls short of it by a few inches, or as much as circumstances will allow.

Butt Joint occurs amongst other positions when circular earthenware, or stoneware flue pipes, or vent linings are built in a stone wall with their ends butting. In tracery, fret work, columniation, &c., the butt joint is of common occurrence, and is frequently strengthened with a dowel.

Caulked Joint.—The joints of reservoir walls are rendered water-tight by driving into them iron cement or other approved substance with the caulking tool and hammer. In building in ironwork with caulkings, as explained in the Smiths' Section, the mason must be careful that it is sufficiently protected not to injure the work by its oxidation, or to disfigure it with stains.

Cement Joint.—When this occurs in masonry, it is generally understood to be made with Portland cement, which is stronger neat than when gauged with sand, which reduces its tensile and compressive power. It is, however, most usually compounded of one or two parts of cement to one part of sand, quite free from impurity. Coping, chain bond, cramps, and dowels are commonly set in it. It is particularly necessary that masonry exposed to the action of the sea should be carried up with quick-setting cement immediately it reaches the mean sea-level, and it has been suggested that, in running water, the stronger but slow-setting Portland might be advantageously pointed with the weaker but quick-setting Roman or black cement. In mending stone, masons use a cement made of two parts of resin and one part of beeswax, which are melted together in a crucible over the fire and then poured into cold water, worked up well together and formed into sticks. The broken surfaces being well heated with a hot iron, the cement is melted and spread over them, and the joint completed by pressing them hard against each other. For soft stones, red shellac dissolved in naphtha, brought to the consistency of thick glue and spread over both surfaces, suffices, but a small dowel previously fitted makes a stronger joint. For inside work, plaster of Paris mixed up thin with a little dust of the stone is often used for the same purpose, the surfaces being first wetted.

Chamfered Joint is found in rustic work when the arrises of the blocks are bevelled off to edges forming an angular groove, as shown in Fig. 12, whose sides stand at right angles to each other.

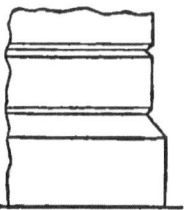

Fig. 12.

Checked Joint.—This is synonymous with rebated joint.

Circular Joint is one in which either circular sunk or spherical work occurs. It is made when he and she joggles are of circular or curved profile, or when the bed joints of arch stones are embossed, &c.

Clamped Joint.—This is another name for cramped joint.

Close Joint.—This always occurs in granite or other kinds of ashlar facing, where it is from $\frac{1}{8}$ in. to $\frac{1}{16}$ in. thick, and in ashlar without rubble backing, where it should be throughout about $\frac{1}{4}$ in. thick, which is sufficient for the strongest work, and little enough to preclude trouble from settlement. To obtain it the beds and joints are dressed true, square, and full the whole depth of the block, or at all events for some distance back, and the chisel drafts round the edges form sharp and straight arrises; but often the stones are only accurately dressed a few inches back, and the rest chipped away and underpinned, leaving insufficient bearing surface. When set properly, the stone is first lowered on its bed before spreading the mortar to try whether the bed and joints are dressed level and true, and if the stone rests with its face in a true plane with that of the other part of the work, or otherwise correctly. Protuberances, if any, are then marked, the block is removed, superfluities are knocked off, beds and joints brushed and wetted, the mortar bed is evenly and thinly spread with fine mortar to within an inch of the front arris, to leave room for a strip of putty as described under Fine Joint, and then the block once more descends to be well and truly laid and settled in place through the active agency of the mallet. A close joint can be obtained with roughly hewn beds, provided they are carefully and regularly axed or broached, and the front arris chisel dressed. The stone must be set on a continuous cushion of fine mortar or cement, without void or packing, showing a perfectly even thickness on the face, but little in excess of the average thickness throughout the bed. When close joints are thus obtained by dint of careful workmanship, it is another matter to preserve them, and where violent shocks have to be withstood the precautions described under Bolted Joint must be observed. Engineers wisely insist that all masonry, whether ashlar or rubble, exposed to the fury of the waves, should be set with full close-fitting joints, or at all events with joints completely and solidly filled in with good hydraulic mortar, as quick setting as possible compatible with the requisite strength. By such means the evil effects of syphonic action arising from water being forced into the joints are avoided. From these observations it will be seen that weirs require very close joints.

Coarse Joint.—A maximum thickness of from $\frac{3}{16}$ in. to $\frac{1}{4}$ in. is amply sufficient in coursed ashlar for strength and durability, presuming that the beds and joints are properly dressed and smoothed. Block-in-course masonry, with roughly dressed beds and joints, requires a minimum thickness of $\frac{3}{16}$ in., consequently a coarse joint in this case exceeds in thickness what would be classed as such in ashlar. All uncoursed rubble masonry is built with joints of irregular thickness, and comparatively much coarser than those above noticed, and the mortar, therefore, of which so much is required, should be of the best possible description. Provided this is the case, and that a pretty average thickness is preserved throughout the work, coarse joints are certainly more reliable than thinner ones permeating a superior class of masonry where the fundamental requirements of properly tooled and levelled beds and joints are merely observed a few inches back from the surface.

Coursing Joint is the same as described in the Bricklayers' Section.

Cramp or Cramped Joint occurs in masonry when two blocks are locked together with a metal or slate cramp. If of metal, it should preferably be of copper or bronze rather than of iron, and cast or galvanised iron rather than wrought. A metal cramp consists of a bar with turned-down jagged ends, or else of double dovetail form, in the latter case sometimes 15 in. or 18 in. long, 4 in. broad, and 1 in. thick, fitting into a groove just large enough to admit of its being enveloped in Portland cement, sulphur, or lead, and left flush. Plaster of Paris is used for securing cramps in interior ornamental marble work. Slate cramps are cut to fit tightly, or they may be run like the others. The main difference nowadays between cramps and dowels in masonry may be said to consist in the former being always let in on the top surface of two stones of the same course, to hold them tight at the joint and not at the beds. Chimney pieces and other marble work are put together with small copper cramps. Small iron cramps made out of $\frac{1}{4}$ in. galvanised iron wire are likewise used. There is always a risk attending the use of iron cramps in stonework, both as regards stains and dislocations, as noticed under Run Joint, no matter how well protected from oxidation they may be, for some unforeseen contingency may abrade or displace part of their protective coating; yet at the same time if the exclusion of air is perfect there can be no corrosion. With any system of metal cramps whatever a good lightning conductor is imperatively necessary, owing to its determining instead of the cramps the polarisation of the electric fluid just previous to a discharge. Chain bars, mentioned under Collar Joint in the Smiths' Section, may be viewed as expanded cramps, the bars being furnished with stubs corresponding to the fangs or corkings of a cramp.

Cross Joint.—A transverse as opposed to a longitudinal joint. In ashlar the term "side joints" is more usual than cross joints, but the single word "joints" signifies, and is almost always used to signify, those surfaces of the blocks which form the cross joints. Face, back, beds, and joints comprise the six surfaces of an ashlar.

Dovetail Joint occurs in masonry when part of one stone is cut in the form of a reversed wedge to let into a sinking in another stone cut to receive it. Or it may be only cut as headers in ashlar facing are, viz., wider at the tail and for better bonding with the backing. In dovetailing together blocks in a course of ashlar, it is usual to sink the sides of the blocks so that the space between any two may form the expanding and contracting hole for a similarly fashioned block to fit into. Hence the transverse joints run in a zigzag direction as shown in Fig. 13, instead of taking the usual rectangular course, and in a similar manner the courses may be dovetailed together as well as the blocks.

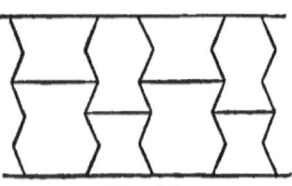

Fig. 13.

Dowelled Joint.—This is a joint fortified against lateral motion by inserting into opposite mortise holes in abutting stones a pin or plug of simple section and hard durable material, and of a size varying more or less with the magnitude of the stones. This is called a dowel, and it is useful in arch work, in the facing of breakwaters, piers, quay and dock walls, piers and abutments of bridges, &c.,

and is sometimes inserted at every joint. It may be a piece of copper or iron of double dovetail form, but more frequently consists of a slate cube from 4 in. to 1 in. square. Cast iron dowels about 1½ in. thick and double the length are used for the drums of columns, but not so frequently as small slate cubes. For marble work generally copper dowels 2 in. long and ½ in. square are appropriate enough. In all cases the dowel ought to be fairly set and made to fit tightly either by friction or by being fully run with cement, lead, or sulphur, &c. When occurring between the beds of ashlar, a dowel is sometimes denominated a bed-plug and sometimes a bed-dowel-joggle.

Elbow Joint occurs when a crosette is formed, or when a voussoir and part of a horizontal course are cut out of one block.

End Joint.—This is another name for side joint and also for heading joint in arch work, coping, kerbing, &c.

Face Joint occurs in the face or front, or most important elevation or surface of a structure. It varies greatly in specific designation according to the finish of the visible part, which, for instance, may be either close, chamfered, or pointed. The term is perhaps most frequently used in connection with the face of an arch. In a skew arch of equal thickness from crown to springing the face joints are slightly curved and their chords radiate to a point below the axis.

Fair Joint signifies in face work a joint in the same plane as the rest of the work, both true and square, and generally speaking one executed in a neat, exact, and workmanlike manner, with parallel surfaces out of winding and carefully dressed.

Fair Axed Joint occurs in granite when the beds, joints, and arrises are worked fair, that is to say, true, square, and straight, the stone being axed as well as broached, *i.e.*, dressed fine and close, with the pick or else with a hammer and punch previous to being hewn to a nearly smooth face with the axe. The pick is a heavy hammer with pointed steel tips, and brings the surface of granite simply out of twist or to a fine finish, but not so fine as that wrought by the granite or patent axe, or patent hammer, which, possessing a head consisting of six or more steel cutting edges, about 3 in. in extent, can put the finest possible axing, or dressing, or finish upon it. Granite, to be worked well, requires specially trained masons.

False Joint is made by running a groove to imitate a joint.

Filleted Joint.—Fillets of stone may be built up with the wall to receive the slating, as described under the corresponding bricklayers' joint.

Fine Joint.—The fine joints of ashlar are usually formed with a putty composed of lime, white lead, and a little fine sand, the fine mortar of the joint being kept back about an inch from the face to make room for the strip, which not unfrequently, however, consists of plasterer's putty or even common glazier's putty. Thin plates or pieces of lead are sometimes used to obtain fine joints as well as a little desirable play in providing a bearing where there has been some uneasiness as to variable pressure in the case of a voussoir or column tambour. A change from very thick to thin joints took place in the eleventh century.

Fine Tooled Joint is made by finishing off the margin drafts, worked round the stone smooth and neat, with the chisel termed a tool, which has a cutting edge about $3\frac{1}{2}$ in. broad, the remainder of the beds and joints being sufficiently dressed by knocking off the protuberances to allow the drafts to come close. Or the whole surface of the joint is tooled smooth after being boasted, the marks left by the boaster, which may be either smooth and regular or random and rough, being taken out by the tool.

Flushed Joint occurs when spalls, chips, or flakes are

splintered off at the arris from a concentration of weight owing to the bed being worked hollow. It is also sometimes produced by the use of fat lime mortar, which only sets where exposed to the air, and consequently offers more resistance to pressure at the arrises than further back from the face. The same term obtains when the inside joints are filled up with mortar and brought up to a level flush with the top surface of the course without leaving vacuities, and before the mortar for the next course is spread. The process occurs in all descriptions of coursed masonry, squared rubble, or even random rubble if need be, for the latter admits of being every now and again brought up to level courses and flushed. In stone walling sometimes the ordinary mortar is mixed with one-third fine clean gravel for flushing purposes.

Full Joint.—This will be found described in the Bricklayers' Section.

Grooved Joint is formed when opposite grooves are sunk in adjacent blocks for the passage of rod bolts, or in the end joints of the stones in horizontal and raking cornices to receive white lead. Vertical grooves of some size are occasionally made in facing blocks for the reception of concrete backing. In constructing sea walls without cofferdams, the edges of granite slabs are sometimes grooved to embrace and enclose the projecting flanges of iron piles, the grooved joint being filled in from the top with cement. Concrete and rubble form a good backing to this arrangement, the rubble protecting the concrete until it is set, or sheet piling may be used instead. Iron piles can also be protected and surrounded, and a strong interlocking facing obtained by cutting grooves on the sides of blocks as well as sinking central holes through them by which they may be dropped over the piles. Slabs of stone are correspondingly cut to fit the grooves of the casing blocks, and being slid down into place the grooved joints are gradually rammed tight with cement.

Grooved and Tongued Joint.—This term is given to a joint between two stones connected by a long he and she joggle, or else by a groove and gunmetal tongue. There are no such alternative names as distinguish the similarly styled joiner's joint.

Grouted Joint is formed by pouring thin semi-liquid mortar or cement into a joint to fill up accidental voids and make all solid. The practice has its advocates and opponents, and no doubt if there are facilities for the superabundance of water to run off without washing away or diluting other setting mortar in its course, it may have its advantages in tending to the consolidation of the work. On the other hand, its pernicious effects must be great if the lime be drowned. Grout is requisite in deep and thin joints between large stones, but it is deficient in strength, and should never be used except where thicker mortar cannot penetrate, for too much water deprives lime of two-fifths of its strength. In some cases grouting has been advantageously poured upon the work through a long funnel, acquiring thereby the penetrative power due to a head of liquid as well as to gravitation. A key stone is more likely to have full joints by being fixed with cement grouting, and in ashlar coping to pier walls, flooring to locks or docks, and in many other parts of masonry grouting, especially if of Portland cement, is clearly a source of strength. Cement grouting between the drums of columns has already been described under Bed Joint.

Heading Joint in arch work occurs between two stones of the same string course, and therefore vertically divides the ring courses. In large arches these joints should overlap at least 12 in. In a cylindrical skew arch they constitute a spiral surface, twisting round the axis and intersecting at right angles the spirals of the coursing or bed joints.

Hollow Joint is necessary under window sills, as

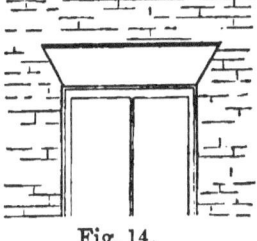
Fig. 14.

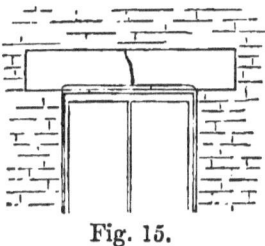
Fig. 15.

described in the Bricklayers' Section. It also often occurs over splayed window heads, as shown in Fig. 14, owing to their subsidence, which is, however, far preferable to their fracture when bedded with square ends and joints, as in Fig. 15. To obviate this unsightly and dangerous misfortune, a flat bar of wrought iron may be worked in so as to relieve the head of some of the effects of its own weight.

Horizontal Joint.—This is a term for level bed, bed joint, or coursing joint when not an arch joint. In spire or other battering work the term is opposed to inclined, sloping, or raking bed joint.

Housed Joint is the outcome of the operation called housing, which consists in cutting or hollowing out on the side or face of one piece of stone a sunk space exactly similar in shape to the profile of the undiminished end of the other piece which is to be let or housed into it. It does not signify whether the stones are moulded or not. Generally all parts of ornamental masonry which abut against projections are thus housed into them. The ends of stone steps are sometimes housed in a wall as much as 9 in.

Joggle Joint.—This term is applied to a square, semi-circular, angular, or otherwise shaped tongue and groove joint generally of equal depth the full way through, excepting at the face, near which it abruptly stops in order that it may be concealed. It is the common method by which the sides or ends of stones, or parts of them, are united in balconies, architraves, landings, &c., and is shown in Fig. 16, the joint being run with lead, or flushed up with cement or plaster. The same appellation is likewise given to any joint in which a stone or metal dowel is used, and such dowel is occasionally termed a dowel-joggle. In some cases the joggle is formed by cutting equal and opposite jogs or notches in two stones, filling the cavity with pebbles, and running with cement.

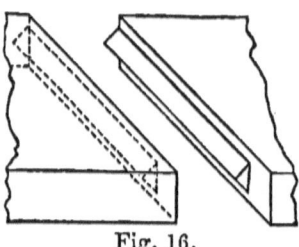

Fig. 16.

Lap or Lapped Joint is formed whenever one end, edge, or side overlaps another, and is, as has been previously observed, the very essence of bond. Ashlar masonry, worked with alternate headers and stretchers, should be well bonded and lapped. Voussoirs likewise. Stone steps, when not rebated, usually form a lap joint with one another 2 in. deep.

Lead or Leaded Joint.—A piece of 5 lb. or thicker milled lead has been often used in columns and arch work, the size of the bed joint, less a width of about 2 in. from the outside, and in mullions, &c., when the beds are very small, to within half an inch from the same. The idea is that the lead yields to the irregularities of the beds, and conduces to a more even distribution of pressure. In turning the arch of the Grosvenor Bridge, Chester, a good deal of lead was skilfully used between the arch stones in plates, wedges, and strips to guard against the joints opening. There is, however, no compulsion as to its use, and the practice is not to be recommended, simply because the same quality which enables the lead to accommodate itself to one alteration of form of an arch ring may, when not cleverly disposed, and under some unexpected concatenation of circumstances, enable the arch to assume another form more at variance with the curve of equilibrium. The same considerations apply with equal force to stone columns.

Mason's Joint is formed by leaving an angular projection to the mortar similar in section to the letter V. The name of this letter is in consequence sometimes applied to the joint, which requires to be made with good mortar to be durable.

Mastic Joint is made by using mastic for the cementing medium. Mastic is an oleaginous cement composed of linseed oil, powdered brick, burnt clay, or limestone, and litharge, the latter constituting about six or seven per cent. of the mixture. It is not so durable as Portland cement, to which it has almost quite given way, but it admits of being painted as soon as applied. Blocks of hewn stone used for ridging were formerly sometimes set with it.

Mitre Joint is the line or plane of union bisecting the angle of junction of blocks, mouldings, &c., when meeting at any angle less than two right angles. In masonry, mitres are scarcely ever permitted between planes, and are never allowed at quoins or angles excepting in mere decorative features, and the mouldings of these are usually either housed or returned on the same block, with an occasional exception, as, for instance, when window sills are returned and mitred to strings.

Mortar Joint.—The remarks in the Bricklayers' Section under this head are more or less pertinent to this trade also. Besides these and others to be found under Bed Joint and Flushed Joint of this section, it may be observed that for ashlar the mortar must be carefully prepared so as to be quite free from grit, $\frac{3}{16}$ in. being a good average thickness, whilst for ashlar facing, or bastard ashlar, $\frac{1}{16}$ in. is sometimes deemed almost too thick. The face of the joint is lipped with plasterer's putty, or, as described under Fine Joint, for a distance back varying from $\frac{3}{4}$ in. to $2\frac{1}{2}$ in.; but some prefer using a fine variety of mortar coloured to match the stonework. In the present day it is usual not to have projecting joints, but to leave the mortar within the surface of the work. For rubble masonry the mortar should be of excellent quality, owing to its forming so large a proportion of the walling. Where it is compulsory to use so much, poor indeed must be the work that is bolstered up by an impoverished quality, for obviously within reasonable bounds the worse the mortar the less of it the better.

Mortise Joint.—Whenever a hole is cut in masonry to receive the corkings or tenons of iron columns, stanchions, standards, door-posts, or ends of balusters, railings, &c., the resulting junction is correctly described as a mortise joint.

Mortise and Tenon Joint.—Formerly this term held for a projection left in one stone to fill up a vacuity made for it in another, but it is now not common.

Notch or Notched Joint is one formed by sinking or cutting indentations in slabs, flags, landings, or blocks to make them fit into their allotted places. Slabs or front hearths are notched for chimney-pieces. Paving is notched to shape where necessary. Ashlar may be notched to receive arched flooring, transoms, &c., as in Fig. 17. Stones in rubble walling are notched as occasion requires with fair ends against ashlar quoins.

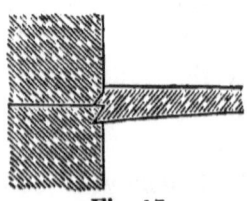
Fig. 17.

Open Joint.—An air-space or open joint is often produced, amongst other causes, by part of the foundation yielding—which usually manifests itself by a crack or "thread"—or by projections being left on the beds beyond the drafts. Open joints exposed to the action of waves of translation may cause much havoc. In arch work, another form of danger arising from a similar source has been indicated under Arch Joint. There are but two ways of curing this evil. If the opening is slight, and only traverses the work like a thread and does not widen, it may be fresh pointed and concealed, but where the crack ominously gapes and gradually extends, all half measures are useless, and the work must be pulled down and the foundation rendered firm and unyielding.

Perpendicular Joint.—This is the same as in bricklayers' work. There are one or two exceptions, however, inasmuch as external architraves and other stone dressings introduce a long description of perpendicular joint which should be concealed as much as possible by the overlap of mouldings and rebates, as noticed under Back Rebate Joint. The drums or shafts of columns should have no vertical joints except from economical motives. In ashlar the perpendicular joints on the face should be exactly over each other, and in coursed rubble the lap of one stone over another should never be less than 3 in. The apex or saddle stones of pediments, gables, and raking cornices should have no perpendicular joint, whilst some allege that a central one is characteristic of the Gothic arch.

Pinned Joint is made by pinning in, which consists in letting into a hole either expressly cut or left, or otherwise provided, the end of a step, beam, or girder, or of almost any piece of iron, slate, stone, &c., and solidly and firmly securing it therein by tightly filling the joint all round with cement, and working and jamming in pieces of tile, old iron, slate, &c., so as to produce thorough compactness and make the joint a source of strength, if that were possible. Stone landings and hanging steps are thus pinned in to a depth ranging from 4½ in. to 14 in., according to the thickness of the wall.

Plugged Joint occurs when a side joint is secured against working loose by means of a plug, often consisting of lead, which is run through a vertical groove into a double dovetailed plug hole that is partly cut out of each stone, as in Fig. 18. In such positions, plugs ought to reach nearly half way down the stones. There are different shapes given to the hole, the object in all being to enlarge the ends of the plug to stop it from drawing. It also is formed when a bed-plug of hard wood, stone, slate, or metal is inserted in opposite mortise holes, or otherwise, in a raking, level, or radiating bed joint in arches, tracery, mullions, copings, &c., to prevent sliding.

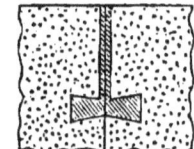

Fig. 18.

This kind of plug should not be less than 1 in. square, but need not be larger than about 3 in. square and 4 in. long when the blocks are of considerable size. In masonry it is no uncommon thing to attach an identical meaning to the terms plug and dowel.

Pointed Joint is effected by finishing off a joint with the point of the trowel in any regular and neat manner. It may be either the original bed joint or else that formed by subsequent raking out and stopping. The term includes making good an open and defective joint, pressing and squeezing in the mortar so as to fill up all cavities, and in some cases even comprehends bedding as well. Rubble walling is often pointed with coal ash or blue mortar. Whenever joints are raked out, they require well brushing and wetting before pointing, and where face stones and cement bond courses are bedded in Portland or other hard-setting cement, they should invariably be raked out clean an inch or so back as soon as the block is laid, or before the cement has had time to set, otherwise there will be a risk of damaging the arrises in pointing, and of not making the face joints preserve an equal thickness.

Putty Joint occurs in ashlar work when the beds and joints are lipped, as explained under Fine Joint and Mortar Joint, with a strip of some kind of putty.

Radiating Joint is one whose direction is obtained from, or whose tendency is towards, a centre or axis. In a right arch it was formerly called a summering by workmen, and in bridges it not unfrequently commences at the foundation or timber floor of the piling, being concealed at times by the outer courses which are laid, for the sake of appearance, with level beds. This kind of joint ought always to be normal to the soffit, but in a skew arch it usually forms part of a spiral surface that twists round the axis of the cylinder, by which means an equivalent degree of strength is obtained. This surface can be easily imagined by supposing it to be described by a straight line travelling uniformly along the axis whilst revolving round it at the same measured rate. Radiating joints occur in quay walls built with a curvilinear batter, and also in circular work with level or inclined beds, in which case an iron rod may with advantage be driven into the ground where feasible as a working centre in case of its being necessary to refer to it to keep the courses to the proper curve. In spire ashlar with raking beds the joints are little benefitted by radiating towards the axis, owing to the magnitude of the angle of repose of dry masonry. Joints in tracery radiate to the centre of the arch, or circle, or arc in which they occur. Conical openings in walls, ribbed groins, domes, niches in circular walls, &c., have radiating joints. The back rebates of stone steps likewise acquire increased stability by being cut to radiate.

Raglet Joint.—This is made in fixing a metal apron, &c.,

when there is no convenient mortar joint into which the turned-down edge thereof may be inserted. The mason cuts a raglet, or groove, or chasing for the plumber to insert therein and secure the edge, the raglet being about an inch in depth and sufficiently wide to take the edge of the metal and the lead wedges, or bats, or molten lead required to fix it.

Raised Joint is one brought forward from the face of the work, and is usually formed by raking, stopping, and pointing, so as to make the joint prominent with a flat, beaded, or angular finish. Unless very well done with good lime it is apt to split off in course of time, especially if the stopping was negligently suffered to be too dry when laying on the pointing stuff.

Raked Joint.—This is the same as described in the Bricklayers' Section, and is much practised in rubble masonry, so far as regards clearing out the joints prior to stopping them with a superior mortar to that with which the stones are set.

Rebated Joint is formed by sinking a rectangular one-sided groove along the edge of a block for doorcases, &c., similar to that described under Back Rebate, or to the rebate taken by the joiner out of a board. In the same way holes in sink stones and sinks for traps, and in stones for coal plates, &c., are rebated to make a suitable and efficient joint, either fixed or movable, as the case may require. The exterior facing of breakwaters built on the vertical system sometimes consists of granite, or beton, or other blocks, checked or rebated into each other.

Recessed Joint.—A rustic joint, described below, being grooved or channelled affords a good example of what may be done by recessing the joints and making them a bold characteristic feature of the work. The only valid reason why mortar joints in masonry are not more frequently sunk or recessed is that a certain amount of weathering is lost by the process, the question of economy being altogether swallowed up by that of effect.

Rough-tooled Joint occurs between surfaces roughly picked or axed, or rough chiselled with the broad axe or boaster, by which they are batted over at a constant inclination to the edge of the stone, or the surfaces may be dressed with the ridges and channels left at irregular distances. In mediæval times the bed joints of the drums of columns were frequently left rough-tooled, and a uniform bearing obtained by running them with lead. Custom varies now, however, in different localities, but usually the above or a similar kind of dressing is confined to the margin drafts, the superfluous stone within them being simply knocked off until all parts of the surface approximately coincide with a banker rule laid upon it. The kind of work bestowed on beds and joints generally depends on that on the faces. If these are polished, then there is an inclination to introduce lead into the joints, or else they are worked so smoothly as to endanger the loss of the mortar hold. Ashlar beds and joints are usually drafted or boasted, or, in other words, dressed with a chisel called a boaster, 2 inches wide at the cutting edge, which leaves the surface covered with marks like ribands or small chequers at regular distances apart, the drafts being tooled very neatly to allow the stones almost to close at those parts.

Rubbed Joint is a polished, cleansed, or smooth joint, and is described under the latter term.

Run Joint.—The stones of cornices and pediments are sometimes run with white lead, that is to say, white lead made into a semi-liquid consistency with linseed oil is poured into their joints. Slate dowels and joggle joints are similarly run with Portland cement grout. There can be no doubt that grout is useful in thus treating the joints of key stones, &c., where the mortar cannot be well and evenly spread, but its indiscriminate use tends to loosen rather than bind and consolidate. All run joints in masonry should be fully run and well filled in. Copper or bronze ties, cramps, plugs, and dowels are preferable to iron, but involve additional expense, though cast iron, with its original skin intact, is little deteriorated by rust, and is often employed in good work for cramping and dowelling. Neat Portland cement of undeniable quality is perhaps the best material in which to fix and embed iron. Wrought-iron ties, however, bedded in masonry—and often of superlative use—should be invariably protected with a coating of zinc or pitch, or some other anti-corrosive. There should be no excuse for concealing this invaluable metal in an unprotected state in the heart of a stone wall to work havoc by corrosion or disfigurement by discoloration.

Rustic or Rusticated Joint is an ornamental finish given to a face joint, particularly in basements of important buildings, banded architraves, arch work, &c. It is effected by sinking, or grooving, or channelling the joint by chamfering, rebating, or moulding the arrises, as shown respectively in Figs. 12, 19, and 20. The chamfered joint forms a right angle, and that which is rebated is called a square joint. Sometimes the grooves or rustications are restricted to the horizontal joints, but as often as not vertical grooves accompany the horizontal. In minor buildings quoins, either plain or rockworked, show rusticated joints. The width of the grooving is quite a matter of taste. Columns not unfrequently have deep, thin, and square rustics, for obviating or more or less concealing flushed joints or edges.

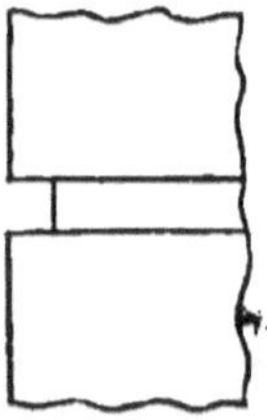

Fig. 19.

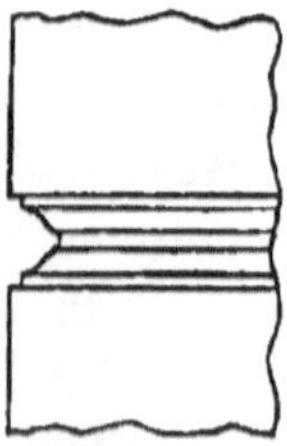

Fig. 20.

Saddle Joint is the outcome of the operation known as saddling the joint, which consists in not weathering the stones of a cornice or coping quite up to the joints, but leaving them at their full height at each end, so as to form a small weathered ridge or saddle to throw off the rain, as shown in Fig. 21. This kind of finish is further noticed under Vertical Joint. A stone cut with two skewbacks to form a double springer on the

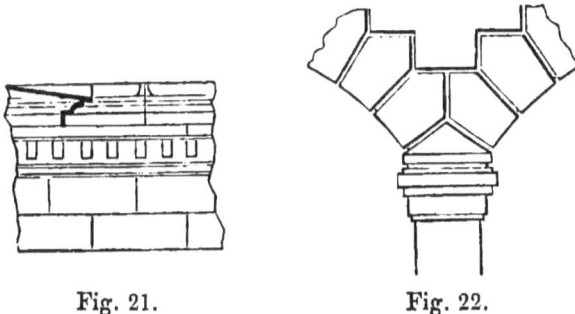

Fig. 21.   Fig. 22.

top of a pier, as in Fig. 22, for the support of two opposite arches, is sometimes called a saddle piece, and the continuous joint between such stone and the overlying masonry is another form of saddle joint.

Sawn Joint occurs in flagging and elsewhere between stones when no other labour besides sawing is executed on the sides or surfaces forming the joint.

Scribed Joint.—This is an exactly similar joint to that of the joiner of identical name, and occurs where stone cornices and mouldings run into and unite with corresponding ornamentation in cast iron, which happens in bridge work and other situations.

Shift Joint is the same as described in the Bricklayers' Section.

Side Joint in ashlar is a vertical cross joint usually simply called joint, and is worked like the bed. In ashlar facing it requires strengthening with copper or gunmetal cramps or dowels. Iron cramps enveloped in tar form a cheap but hazardous substitute, excepting in unimportant work or, perhaps, when cast or galvanised. In coursed rubble masonry the side joints ought to be dressed square and true, or free of twist; and in squared, snecked, or irregular coursed rubble they are usually roughly squared, and show mortar spaces of varying thickness. In random rubble they follow almost any direction, but are so disposed as to break joint against both the upper and lower blocks. Amongst masons the term "side" often gives place to "end."

Smooth Joint is the result of carefully axing, tooling, chiselling, or dressing the surfaces of contact and finally rubbing them with stone, sand, and water, to a degree of exactness and fineness that cannot be surpassed without losing what little key remains for the adherence of mortar.

Square Joint.—One that is perpendicular to the face and bed or axis of the same block, or to the principal face or facial plane of the feature of which the joint is a factor. For instance, in a traceried window all the joints should be square to the axes of their respective bars, transoms, jambs, mullions, &c., and to the window plane as well. To be square, the surface of a joint must be out of winding. A rusticated joint is said to be square when formed with a rectangular groove.

Straight Joint.—This is described in Section II.

Strap or Strapped Joint results from strengthening masonry by building therein wrought iron straps and connecting them by means of rod bolts or tie-bars with screwed ends.

Sunk Joint is of constant occurrence between rebated jambs and frames, rebated holes and plates or traps, and wherever the surface of a stone is in part lowered to form a junction with another.

Tabled Joint consists in cutting a broad and more or less shallow projection termed a table in one stone to interlock in a corresponding recess called an indent in the other. It is a species of joggle joint, a joggle being formed whenever one piece is notched to receive part of another, although it is only in certain instances that the term "joggle" is used to specify the joint.

Thick Joint.—The remarks under Coarse Joint and Mortar Joint in this section, and those under this head in Section II., render further description here unnecessary.

Thin Joint.—This is described under its alternative title of Close Joint. It requires mortar made with fine sand.

Tie Joint is so denominated from its containing some part of an iron tie which usually takes in masonry the form of chain bond, as noticed under Bond, or else of chain bar, which will be found described under Collar Joint in Section VII. Very frequently long wrought iron bars, about 3 ft. $\times \frac{1}{2}$ in., with corkings, are built in at the summit of rubble walls to support the quoins, or to prevent the abutments of an arch from spreading. New masonry is likewise tied to old by building in iron straps along with the new work and attaching them to the old in due course after settlement.

Tooled Joint.—This is formed between surfaces that have been smoothed with the tool or chisel sometimes called the broad axe. After the surface has been boasted or partially or roughly dressed with a chisel called a boaster, about 2 in. wide, which leaves a succession of close parallel marks at one regular acute angle with the edge, it is tooled with another chisel having a cutting edge averaging 4 in. in width, which is passed over the stone at right angles to the edge with slow careful strokes, removing all the marks of the boaster, and leaving instead numerous little ridges and furrows close together and parallel. Joints cannot be too well or too accurately tooled, with this exception that they must not be deprived of that degree of roughness which is necessary to afford a key for the mortar. It is not enough to dress the margin drafts, but where it is considered extravagant to tool the whole surface between them, the dressing should at all events be continued 6 in. or 7 in. back from the face. It is scarcely to be expected that workmen, without close watching, will go on giving the same degree of labour to the surfaces of the block which they know will be lost to sight, as they willingly do to the single and visible one constituting the face; and, as a rule, in ashlar the latter is dressed with a higher finish than the former, though, as has been already insinuated, both proceed as it were *pari passu*, one being often if not always a stage or two behind the other in quality of workmanship.

Transverse Joint.—This is one of the alternative names of a cross joint.

True Joint.—One that is made between two parallel planes, and therefore free from twist or, in other words, out of winding.

Underpinned Joint is made between two blocks when the bed is worked slack towards the back or tail end, producing a joint fine enough at the face but necessitating the upper block to be pinned or propped up at the back part of the bed with a wedge or piece of wood, slate, or stone. It is one of the results of careless or inferior workmanship.

Upright Joint.—This is a vertical or side joint, which in important work is sometimes strengthened with a cast-iron cramp from 6 in. to 9 in. long, let in and flushed and surrounded with cement. At other times the joint is plugged with lead or else secured with a joggle about 2 in. square, filled in with pebbles and cement, but very often no connection whatever beyond the cementing medium is used.

V Joint.—Will be found described under Masons' Joint.

Vertical Joint.—This term includes cross, or side, or end, or transverse as well as back joint; but in ashlar masonry the term "joints" by itself is usually understood to signify the side joints in contradistinction to the bed joints. The whole utility and function of bond concentre in the proper disposition of the vertical joints. They should lap well over those of the course below whilst the perpends are truly maintained. In block-in-course walling, however, which forms a sort of transition style between ashlar and rubble, the perpends are not kept, though a good lap is carefully preserved. Where projecting mouldings occur, as in cornices, strings, lintel and sill courses, &c., it is advisable to weather or saddle the upper edges of the vertical joints, and to set and point them in Portland cement or oil mastic of the same colour as the stone, and in some instances the vertical joints of the main cornice are topped with rolls hewn out of the solid and weathered.

Water Joint is a term given to one formed between the stones of a cornice by not weathering them throughout their whole length, but leaving them level or nearly so for about ¾ in. on each side of the joint.

Weather Joint is one made proof against driving rain and wind, and results from careful bedding and pointing, so that it may be throughout full and solid, and wholly devoid of threads or crevices arising from imperfect adhesion between the block and cementing medium, which perhaps depends as much upon the quality of materials as on that of the workmanship, it being impossible to obtain proper adhesion with bad mortar. All open and defective joints, sills, frames, &c., must be carefully pointed up at the completion of a building, and in carrying up the work the stones must be wetted and the mortar not prematurely checked in setting by the incautious use of grout, otherwise countless minute watercourses will be formed after rain between the sides of the stones and the impoverished mortar. Lime will not reject a superfluity of water like Portland cement, nor will it fix upon what it needs with such certainty, and consequently the mortar, of which it forms so important a factor, should be treated with some regard to the conditions that regulate its setting as well as to the porosity or absorptive power of the stones it is intended to fast bind together. Besides the precautions observed in weathering exposed surfaces and saddling the upper parts of the vertical joints of projecting and unsheltered work, it is necessary to throat and undercut the lower surfaces of strings, labels, &c. All projecting stone sills are throated, that is, provided with a shallow and narrow groove on the under side to prevent the return of the drip, and the white-lead joint between the stone and wood sill is still further protected by the insertion of an iron tongue about $\frac{3}{4}$ in. by $\frac{1}{4}$ in., which is bedded in red cement in opposite grooves cut in the sills.

SECTION IV.

TILERS' JOINTS.

Back Joint.—This corresponds with that of the same name made by the slater.

Bed or Bedding Joint is made by hanging plain or crown tiles on laths with oak or fir pins, or large flat-headed wrought copper and zinc or galvanised iron nails, or by projecting ears, cogs, lugs, or lips, moulded on their top edge. In some cases, both on roofs and walls as weather tiling, tiles are laid and bedded on each other in mortar made with hydraulic lime, to which ox hair has been added. In exposed parts, cement is sometimes preferred. Rolls for ridges and ornamental crests are bedded in cement. Concave tiles for hips are bedded in mortar without lapping, but there are proper hip tiles, however, that are made and shaped to lap, whilst valley tiles, which are somewhat triangular in shape, always do so, and are hung and bedded with pins like the flat one. A T-nail and hip hook dipped in pitch were formerly used with each ridge and hip tile respectively. Tiles are similarly laid on iron roofs, angular iron laths packed with wood forming the bearing surface for fastening upon, but with special-made tiles, as used on the Continent, no wood is required, the tiles reclining upon the laths, to which they are sometimes tied by copper wire passing through drilled holes. Ornamental, glass, and other tiles are treated in much the same fashion as the plain variety, the object being to obtain a dry covering so fixed as not to be stripped by the wind. Bedding tiles in hay or moss is now seldom practised. Pantiles are hung on laths or battens by means of a knob, tongue, or stub under the top edge, with a lap of about 3 or 4 inches between the head of one and the tail of another, there being no intermediate tile as in the case of plain tiling. Wall tiles are sometimes bedded in fine plaster on a backing of Portland cement.

Bond or Bonding Joint.—In roof coverings and weather tiling to walls this is identical with the corresponding joint of the slater. Tiles of fanciful shape, however, requiring peculiar capping, have special modes of bonding.

Broken Joint.—This is made by causing the side joints along each course or margin to fall between those of the contiguous ones. It is indispensable in bond, which is nothing more than regular and systematic lapping.

Cement Joint.—Ridge and hip tiles are bedded in Portland cement. T-nails, sometimes used in each joint to secure the former and hip hooks the latter, both nails and hooks being dipped in hot pitch, are things of the past or quickly becoming so.

Dowelled Joint is formed when ridge tiles are connected by means of oak dowels set in purpose-made holes in the rolls.

Filleted Joint.—Filletings or lutings of cement, cement and sand, lime and cement, or lime and hair plaster, that is, common hydraulic mortar mixed with ox hair, are formed in the roof to make the joint tight where brickwork, &c., cuts into it, and these filletings are keyed by means of cast iron nails driven into the wall pretty closely together, say not more than 3 in. apart.

Flanched Joint is one protected by a weathering or covering, consisting of one or more tiles bedded slantwise over it in mortar or cement. Flanching answers the same purpose as the filleting just described in the previous paragraph, and, in fact, by many the terms are viewed as synonymous.

Heading Joint occurs in ridges between ridge tiles, and is either secured with oak pins, as noticed under Dowelled Joint, or else if executed in mortar with T-nails plunged in pitch.

Key Joint occurs when strong cast iron nails are driven into brickwork sufficiently close, or about 3 in. apart, to form a key to hold filleting.

Lap Joint.—This involves the principle of tiling, and has been already sufficiently alluded to.

Mitre Joint.—If there are no hip tiles, what are called skew tiles are sometimes used and bedded in cement and mitred at the hip. The tiler rarely cuts his tiles.

Mortar Joint is formed when tiles are laid in mortar, which often happens to a 3-in. lap. In iron roofs tiles are laid on wood laths with or without mortar. This joint is further described in the last paragraph of this section.

Pointed Joint occurs when plain tiling is back-pointed on the inside when the pitch is low, or as noticed under Weather Joint. Pan tiling is likewise made as air-tight as possible when it is not intentionally left open for ventilation, by pointing both horizontal and vertical joints on the inside with hair mortar composed of ordinary mortar and ox hair. Sometimes, however, the side joints of both pan and plain tiling are pointed on the outside, but the tails of the latter should not be thus sealed, otherwise the water that elsewhere finds its way through the covering cannot escape.

Weather Joint.—To keep out the weather with a covering of plain or crown tiles, these should be of the best quality, and glazed if possible, the roof pitch being not less than 30°. The tiles must be laid on strong oak or fir laths to a gauge of not more than 4 in. at the most, or to a 2 in. lap or bond. In all exposed situations, or in any with a pitch less than what is called a true pitch, which is 45°, the tiles ought to be bedded and pointed with hair mortar, prepared with hydraulic lime and ox hair. There can be no doubt, however, that with the object of keeping out the snow it is best in all cases to point the inside, taking care to use mortar that will stick. So long as the water is thrown off and the framing made sufficiently strong, there is not much advantage gained in having a high pitch beyond lessening the strain on the members of the principals, but in many instances after a high pitch has been decided on, the increased height thereby obtained for providing additional accommodation has not been utilised, and consequently the extra expenditure has been thrown away. The pitch of a roof is the angle having for its tangent the ratio between the height of the roof and the half span, this determining both the length of the rafter and its inclination to the horizon.

SECTION V.

SLATERS' JOINTS.

Back Joint is the same as that formed by torching or back-pointing, and is described under Torched Joint.

Bed or Bedding Joint.—Slates are bedded on close boarding either rough or wrought on one side, or else on battens about $2\frac{1}{2}$ in. × $\frac{3}{4}$ in., the former being preferred, as it keeps the roof cooler and drier, and forms a stiffer covering. Asphalted felt, which is both damp proof and a non-conductor of heat, is sometimes laid over boarding, and battens occasionally, in good work, again over this. The distance between the battens is the same as the gauge, which is equal to half the difference between the length of the slate and the lap. If the slates are nailed near the centre the full length of the slate is taken as the measurement, but if near the head then one inch is deducted therefrom for the space between the nail holes and the head. Slates are usually kept down on their beds by double nailing without straining or bending them through holes punched either near the heads or centres, the nails used having large flat heads, and being either of copper, zinc, galvanised iron, composition, or iron dipped in boiled oil. Slates are cut and mitred at the hips (over strips of 6 lb. lead lapped and nailed), and cut at the valleys. At the eaves, and where breaks or rakes occur, they are bedded on tilting fillets as a wind protection. Thicker pieces or slate fillets are often used for ridges and hips, being nailed and bedded in slate mastic. In other cases ridging, either of sawn slate, hewn stone, earthenware, or glazed stoneware, is used, and bedded and close jointed in cement so as to drop dry, and such ridging is tied occasionally together by an iron rod passing through the perforated rolls. There is no variation of any moment when bedding slates on iron roofs, the boards and battens being easily secured to wood packing fastened to the iron bearers. Slates can also be bedded in mortar directly upon sloping brick work.

Bond or Bonding Joint.—This is similar in principle to that of the Bricklayer, and is obtained by making the side joints of each course alternate with those of the courses above and below, a double course being laid at the eaves.

Broken Joint.—This is explained under Bond. It is always necessary in slating, and usually it is enjoined that the joints should be properly broken.

Cement Joint is formed when fixing slate ridging, which is set in oil cement or mastic, and secured with screws, the cement being painted to match the slate.

Cistern Joint.—This consists of a groove coated with red cement, as described under Flange Joint in Section VII., into which another slab is fitted.

Close Joint signifies as close a joint as possible, as, for instance, when hewn ridge stones form the ridging they are close jointed and set with oil mastic so as to be perfectly watertight. Some particulars pertinent to this paragraph will be found under Side Joint.

Close Cut Joint.—This signifies that the slates are not trimmed with too ragged edges, but cleanly cut to fit closely at the doubling eaves course and wherever necessary at hips and ridges, or where dormers, chimneys, ventilators, &c., break into the roof.

Close Cut and Mitred Joint is sometimes formed

at the hips of a roof, and is made by cutting the slates to the proper bevel for mitring, and bedding them either on oil putty or on soakers about 18 in. or 20 in. wide, cut out of 2 lb. or 3 lb. lead, and having similar lap and length to the slates. If the lead is laid on in long strips it should be at least 6 lb. lead. When putty is used, which is of constant occurrence in common work, it is painted over to correspond in colour with the slates.

Filleted Joint is one protected by filleting. The fillets are sometimes narrow strips of slates nailed to the ridge piece of a span roof, or to brickwork in the case of the head of a lean-to roof and the sides of both varieties, if abutting thereon, and pointed in oil cement or mastic, or Portland cement, or else they consist only of Portland cement, or hair mortar, or even common mortar laid against the wall on cast iron nails every two or three inches apart. Equal parts of cement and sand, gauged with a little mortar, forms a good material, and the work being wetted, the fillet is run about 2 in. up the brickwork and 1 in. over the slates. A cool day should be chosen for the operation. These fillets are left plain or marked with trowel creases. Gables are pointed or run with cement fillets under the slates.

Heading Joint occurs in slate ridging, and is either a butt, but more properly a rebate, in which case it should be cemented with red cement. It is sometimes additionally protected by being tongued with a strip of copper.

Key Joint is made similarly to the corresponding joint of the tiler.

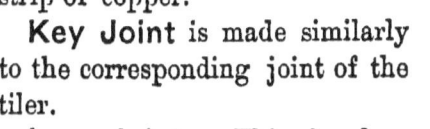

Fig. 23.

Lap Joint. — This involves well-nigh the whole theory and practice of slating, systematic and thorough lapping being the only safeguard

E

against the intrusion of wet. What is technically called the bond or lap is the distance between the nail holes in the head of a slate and the tail of the next but one above it, as shown in Fig. 23, where A B indicates the lap, which should never be less than 3 in. When the nailing is effected near the centre of the slates, the lap is measured from the head, which is assumed to be 1 in. beyond the nail holes as punched for nailing near the head.

Mastic Joint.—Slater's mastic is usually either putty or white lead, and is employed, amongst other purposes, for the joints of the thick pieces of slate sometimes used as coverings to hips and ridges, in which positions they are properly squared and screwed down to the boarding with copper screws. The mastic is afterwards painted to match the slating.

Mortar Joint.—Slates are bedded in mortar on boarding to obtain a sound bearing if much running about over them is expected. This process is further noticed under Shouldered Joint. If rendered on the under side when laid on battens, the mortar should be haired and coloured with smiths' ashes or coal-dust. The same kind of lime and hair mortar is likewise preferable for filletings when adopted in lieu of flashings.

Open Joint occurs in ventilated or open or half-slating, which is employed for covering sheds and similar buildings where a perfectly waterproof roof is not required. It consists in laying the slates with their side or vertical joints open to the extent of from $1\frac{1}{2}$ in. to 2 in., or, in other words, placing the slates that distance apart, so that the smaller the slates the greater will be the aggregate space left partially open.

Painted Joint occurs frequently in common work, being a finish given to conceal the oil putty with which the joints at the hips and ridges, in the cheeks of dormers, &c., have been stopped.

Pointed Joint.—To make slating air and snow tight it is sometimes pointed on the inside with hair mortar. When this is laid all over the slates between the laths or battens, and well tucked and pressed in over the rafters so as to completely fill up all cavities between the wood work and slating throughout, it is sometimes otherwise called rendering. The vertical joints of weather slating to walls, dormers, &c., are pointed with hair mortar and left white, or with oil putty and painted.

Rebated Joint is made between slabs by cutting out equal and rectangular grooves along both edges to be united and fitting them together, as explained under the corresponding joint in Section VIII. In roofing, the slabs are rebated and screwed down to the rafters, the joints being stopped with oil cement or putty and covered with slate half-rounds, bedded on the same material and screwed; but settlement often causes such a covering to constantly leak, and though its appearance is satisfactory, it should not be used excepting when it is properly supported by trusses. Slate rolls on hips and ridges require rebated heading joints, unless the rolls break joint on both sides over the wings, saddles, or flanges of the ridging.

Shouldered Joint.—A modified form of bed joint resulting from the process called shouldering, which consists in bedding the heads of slates on fillets or shoulders of haired lime or mortar about 2 in. in width to make them weather-tight and close at the tails, and less liable to rattle or to crack when stepped on. The mortar may be brought to the same colour as the slates by incorporating ashes therewith, and in some instances the fillet, when on boarding, is developed and extended into a bed for the entire slate to repose on. Where this is the case, the process is more appropriately, though somewhat inexactly, called rendering, and the bed joint becomes a mortar one.

Side Joint is one between the edges of slates, which

are always left more or less ragged and bevelled after being trimmed. Some slaters prefer laying slates not quite close, but a trifling distance, say $\frac{1}{8}$ in. to $\frac{1}{4}$ in., apart, for the purpose of leaving a channel for the easier departure of the water, and this practice is more especially regarded with favour where ton slates are used. This joint is often called vertical joint, and should always be made to range with others in alternate courses throughout the whole slope, that is, from eaves to ridge.

Torched Joint is produced by the operation of torching, which consists in back-pointing with hair mortar slates that are laid dry in the customary manner. The mortar is applied from the inside along the top edge of each batten, and is instrumental in keeping out drifting snow, but it is not very durable in its perfect state, owing to the loosening effects of alternate cold and heat.

Vertical Joint.—This is synonymous with side joint.

Weather Joint.—In order that the joints may be weatherproof, the pitch of a roof for a covering of ordinary slates must not be less than about $26\frac{1}{4}°$, the pitch being sometimes regarded as the ratio between the height and span, the double of which is the tangent of the angle of pitch. Thus, for example, a roof is said to be one-quarter pitch when its rise is one-fourth of the span. Equally often the amount of slope or the angle of inclination of the rafter to the horizon is denominated the pitch, which is high or steep, flat or low, according as it much exceeds or falls short of the so-called true pitch of $45°$. It is quite as important that the pitch should not be so steep as to shoot off heavy rain with too much velocity for the gutters to properly discharge, as it is to avoid too much flatness, whereby the wind may drive the dripping rain under the tails of the slates or even lift them up by its violence and blow them off. As a rule, slates are most likely to keep out the weather when laid on close or open jointed sarking or rough boarding (which is better perhaps nailed diagonally), with felt between it and the slates, boards being stiffer than battens. Sometimes, however, the latter are used as well and over the former. Heavy slates are preferable for exposed situations, and these may be laid with as flat a pitch as 22°. Light sky blue ringing slates are the least absorptive. The bond or lap should not be less than 3 in., and the slates should be secured with two stout copper nails, with a double course at the eaves, no slate being laid lengthwise. Slating nails made of best quality V. M. zinc, or of composition, iron wire, &c., are much used, but when of iron they should be painted or dipped in boiled oil. The nails vary in length from 1 in. to 2 in., with two or three intermediate sizes. It is necessary to insist that the slater makes good all damage occasioned by other workmen, especially the plumber, in order that the roof may be left sound and perfect at the conclusion or on the delivering up of the works.

SECTION VI.

CARPENTERS' JOINTS.

Abutting Joint.—This is common in carpentry, and occurs whenever the end of one timber or piece is adjusted to bear upon any part of another, either for support or for prolongation or continuation. The end may be either cut or left square, and may simply rest upon the other piece, or be framed, or in any way joined to it. The junctions of different lengths of wall plates, of principal rafters with tie-beams and posts, and of binders with girders afford familiar examples, whilst the ends of shores and rakers resting against needles at one extremity and the footing piece at the other equally well exhibit the same kind of joint at the surfaces pressed by the abutments. It is worth observing that the end of a piece of timber when cut and shaped or otherwise arranged to fit against another piece to form a joint is called its abutment, and since the direction of the action of butting timbers upon each other is perpendicular to their abutments, these should be cut as nearly as circumstances will permit at right angles to the lines of pressure. The form of abutment, however, affects only the distribution of strain about the bearing surfaces and the strength of the joint, and not the intensity of the pressures transmitted through and along the timbers.

Adjustable Joint is usually formed with gibs and

cotters, as explained further on, for the purpose of tightening up the tie-beam to the king-post after the roof has taken its bearing. Sometimes the principal rafters are similarly united to the head of the post to bring the members of the truss close up to their seatings after settlement and shrinkage.

Angle Joint is formed when the ends of two timbers or pieces meet at any angle with or without interpenetration. A butt and mitre between a strut and straining piece, or block, as in Fig. 63, p. 107, forms a good example. In old half-timbered houses, instead of joining the sills, and thereby weakening the corners with angle joints, angle posts were let into the ground to receive the tenons of the sills, and a short raking strut inserted 'twixt post and sill to take the strain off the tenon, as shown in Fig. 52, p. 97. In log huts, angle joints are formed by notching, as in Fig. 24, or else by halving and pinning. If squared, the logs may be dovetailed or halved.

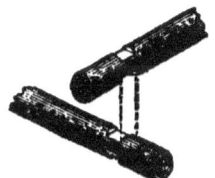

Fig. 24.

Angular Joint is one requiring one of the pieces between which it occurs to be cut to an angle, and is illustrated by a bridle joint or a joggle, or a birdsmouth as represented in Fig. 25.

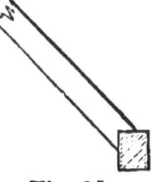

Fig. 25.

Angular Grooved and Tongued Joint.—This is a similar joint on a larger scale to that described in Section VIII. It is used sometimes to help to render sheet piling water-tight when the piles take the form of planking about 4 in. thick.

Banded Joint.—One strengthened by a wrought iron band or strap, usually in duplicate on opposite sides, and bolted through together. Angle bands between struts and beams sometimes cause fracture through the strain being transferred from the timbers to the bolts, and acting with the leverage of one of the timbers, and frequently with that of half the beam.

Bed or Bedding Joint.—Owing to the universality of this joint it will be possible only to allude to one or two of its varieties. The ends of flitch girders and other large beams are bedded on stone templates to spread the weight, with or without asphalted felt or sheet lead as a bearing. When the beam acts as a tie, iron caulking plates or shoe plates are necessary to unite the ends to the templates. A small arch or cover stone will assist in preventing settlement from shrinkage, and in keeping space for a current of air, which should be provided for all round the end. Other timber girders and the heads and sills of partitions are similarly treated, none resting over openings, and all fixed, where possible, with a clear air-space. Binding joists, if bedded on oak templates, may be bolted thereto, otherwise they may be fitted into the sockets of iron girders provided for them, or into iron stirrups passing over wooden girders, or they may be framed into the latter with a double tusk tenon. Tie-beams, if not cogged down to wall-plates, are sometimes fitted into cast iron tie-beam plates built into the wall and projecting therefrom. Common joists ought to be bedded without packing with chips or wedges, and well fitted and nailed down to the plates. When timber is bedded on metal, it is advisable to interpose a piece of sheet lead, with or without wedges, between the wood and the lead, for an air-space. Whilst timber remains dry its carbon is safe, but moisture induces chemical action and causes its loss. The bearings, therefore, should be kept dry, which may be effected by coating the perfectly-seasoned timber with asphalte, but is best accomplished by an air-space.

Bevel or Bevelled Joint is one in which the plane of the joint is oblique to the fibres of one of the timbers, and parallel to those of the other.

Bevelled Shoulder Joint is a variety of mortise and tenon used in uniting inclined to upright and horizontal pieces, and is made by cutting bevelled shoulders to the tenon on the inclined piece, as shown in Fig. 26, and a corresponding sinking or joggle to receive it on the cheeks of the mortise in the post or beam. By means of the joggle the junc-

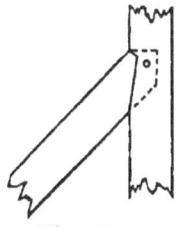

Fig. 26.

tion is made more perfect, and its strength, as well as that of the framing to which it belongs, materially increased.

Birdsmouth.—An angular notch at the end of a common rafter, strut, &c., taken out to enable it to fit snugly down on the arris of another piece upon which it abuts. Fig. 25 represents the usual form of the joint between a rafter and plate, or longitudinal bearer, effected by its aid. Birdsmouths are also advantageously used for the abutment of struts when the latter form acute angles with the rafters or other timbers they help to support, and when no cleats or straining blocks are permissible. There is no reason why a notch, as in Fig. 27, on the edge of a piece should not likewise be called a birdsmouth, since there is but little restriction in the other trades as to its position.

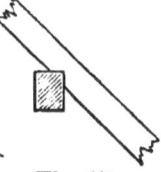

Fig. 27.

Bolted Joint.—This is one which depends either entirely or partially upon one or more screw bolts and nuts for its efficacy and security. A screw bolt, which should be forged from best bar-iron (if not made of copper or gunmetal), usually consists of a headed rod or spindle varying from $\frac{1}{4}$ in. to 3 in. in diameter, but sometimes it is square in part. An ordinary sized bolt in carpentry has an effective diameter of $\frac{7}{8}$ in. or thereabouts; but in deciding upon its stoutness the quality of the iron must be taken into account as well as the description of stress to be withstood, besides which the amount of weakness introduced by boring a large hole through timber must not be lost sight of. At one end of the spindle is a permanent knob called the head or bolt-head, which is, perhaps, all the stronger for being screwed and welded on. The head is preferably square or hexagonal in shape (though it is sometimes cone, rose, or button-shaped), and it equals in width two diameters of the spindle, and in thickness $\frac{7}{8}$ of one in practice; but theoretically the part A B, Fig. 28, need only be equal to $\frac{1}{3}$ the diameter since the section sheared equals A B in breadth by a length corresponding to the circumference of the spindle. The tail-end of the spindle has a screw thread cut upon it, the effective diameter of the bolt being the diameter of the spindle inside the thread; and in order that the cutting of the thread should not weaken the bolt it is sometimes thickened out at the screw end, but as this entails an enlargement of the bolt-hole, the consequence of such a procedure must not be overlooked. The pitch of the screw, that is, the distance between two revolutions of the thread measured parallel to the axis, ought not to be greater than one-fifth of the effective diameter; but it is better for being less, and the projection or depth of the thread is generally about half the pitch. The nut consists of a small block of wrought iron, or other appropriate metal, of square or hexagonal form. It is tapped with a screw to fit the thread of the bolt, and in all cases the clean cutting and proper fitting of the nuts and screws should be well seen to. The thickness of the nut is usually not less than one diameter of the spindle, because the thread lessens considerably the shearing resistance, and its breadth across should equal two diameters. In some cases, instead of a nut, the bolt is tightened up and held by a key or cotter, for which a slot is made in the spindle. The com-

Fig. 28.

bined weight of the head and nut may be approximately assumed as equalling a length corresponding to four diameters of the spindle when the heads are rose-shaped and the nuts square, to five diameters when both are hexagonal, and to six when both are square. In carpentry, washers or discs of metal twice the diameter of the head and half its thickness are interposed at each end of the spindle between the timber and the head or nut, to prevent either of the latter from bruising the fibres. Sometimes the heads when square are let into the timber, which enables the bolt to be screwed up without requiring the head to be at the same time grasped by a tool, but it is a doubtful expedient except on the score of sightliness. In some situations the bolt has to resist tension, and in others shearing, but in either the safe or working strength of a round wrought iron bolt of fair average quality can be found by multiplying the square of its effective diameter by 4 and taking the product as tons. Thus the safe stress that can be put upon a $1\frac{1}{4}$ in. bolt is 5 tons. The carpenter uses bolts and nuts in all kinds of heavy work, such as trusses, partitions, domes, cofferdams, spirelets, framed floors, gates, built ribs, curved and laminated ribs, fished and scarfed beams, flitch beams, trussed girders, bridges, centres, &c., &c. In those instances where the union of parts is entirely dependent upon their holding powers, bolts as used by the carpenter are notoriously unscientific in principle, for they are loosened by the shrinkage of timber, and have no grip upon the fibres like a screw thread. The bolt holes also are a source of weakness, and when racking sets in, the spindles inside and the heads, and nuts, and washers outside crush the fibres against which they bear. They are, however, easily applied, and useful for tightening up work after it has been raised, loaded, and has taken its bearing, but the same advantages are claimed for ordinary wood screws if of sufficient size and turned or driven with a spanner. Before fixing bolts in timber structures, they are sometimes, like other iron fastenings, heated to a blue heat (about 550° Fahr.), and then immediately immersed in raw linseed oil, and painted when dry. It is at all events advantageous that they should be protected from oxidation, and this is perhaps one of the best means to that end, though the immersion in oil is a troublesome and disliked process, and one that is not at all unlikely to be evaded.

Brace Joint is one connecting a brace to the timber or element it helps to confine. As a brace is either a strut or tie, the connection will accordingly be either in compression or tension, and both conditions are respectively noticed under Strut Joint and Tie Joint.

Bridle Joint, sometimes called Notch and Bridle.—This is the name given in carpentry to a joint in which the mortise is supplanted by a tongued notch, and the tenon by a grooved abutment to correspond. The tongue which stands up across the middle of the breadth of the notch, as in Fig. 29, is called the bridle, and is about ½ the breadth in thickness. This joint has the advantage of being more easily got at and inspected throughout all its parts, and consequently of being more likely to be accurately fitted than a mortise and tenon, whose parts are concealed and not unfrequently suspected of being cut without the fulness and nicety essential to an exact fit. Fig. 29 represents a principal rafter about to be united to a tie-beam according to this plan, but a similar kind of joint can be equally well used between upright and horizontal timbers.

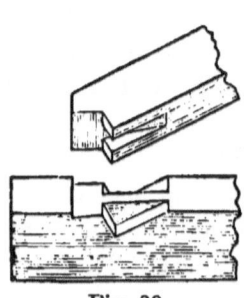

Fig. 29.

Butt or Butting Joint.—This has been generally

explained under Abutting Joint, and it remains only to add that butt refers usually to a square plain joint between two plain surfaces.

Carpenter's Boast.—A facetious name given to a joint in carpentry worked from a centre, and somewhat of the nature of a rule joint as well as that of a dovetail, and useful so far as it accommodates itself in some situations to settlement. It is not considered effective, the reason for which is apparent from a glance at Fig. 30, which shows how prone the joint is to become loose through shrinkage.

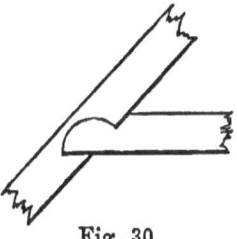

Fig. 30.

Caulked Joint.—There are four varieties of this joint which occur in carpenter's work, though perhaps the first is not actually made by this artificer. It is formed (1) when rendering timber work water-tight by driving tarred spun yarn or some similar stuffing into the joints or seams of planking, sheet piling, &c., and is executed by means of a hammer and caulking iron or chisel, the threads being driven in one after another into the seams, which are previously opened out or reamed with the chisel. After the yarn is well compressed and compacted into a dense watertight mass, it is payed over with a coating of hot pitch. Caissons, bridge timber-platforms, cofferdams, &c., are thus generally caulked. To this class may be added those instances in which pipe sockets are filled with wood wedges and the joints in cast iron tanks made sound with wood slips. (2.) When letting the corkings or caulkings of ironwork into timber, as, for example, when securing with coach screws or staples, the caulked ends of twisted dog irons to joists and wall-plates respectively. (3.) In connecting tie-beams with wall-plates, &c. This kind is made by cutting in the lower timber a notch with a cog left across it, as in Fig. 31, flush with the top surface, and in

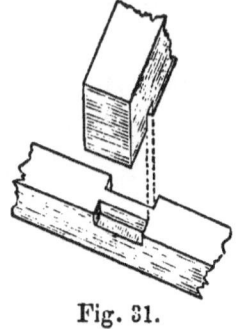

Fig. 31.

the under surface of the upper timber a notch to correspond with, and exactly fit, this cog, by which means the two are prevented from drawing away from each other. Odd as it may seem, this form of junction is called with apparent reason either a caulked, cauked, calked, cocked, cogged, or corked joint. (4.) This variety not only enjoys, according to association or taste, the whole string of aliases just enumerated, but in addition the term keyed. It occurs in built beams, and consists in cutting shallow transverse notches at regular intervals on the upper and lower edges respectively of the two timbers about to be united, which are then bolted through or strapped together with the notches opposite, and when thus well secured they are kept from sliding by driving pairs of cogs or folding keys through the holes so left between them.

Chase Mortise Joint.—This is formed by cutting a chase or mortise having elongated and diminishing cheeks, so that the tenon, on being inserted obliquely into it, can be slid or driven home sideways where there is not room to get it in endways.

Checked Joint is another name for a rebated joint, though the term checked is sometimes used with more extended signification, standing occasionally, for instance, for a groove.

Circular Joint occurs in framing where the abutment of the butting timber is rounded off or circular. It is, however, rarely met with, and there are opposite notions as to its theoretical or practical value. It is otherwise termed Rounded Joint.

Close Joint.—This is not so fine a joint as the joiner's of the same name. It occurs between gutter boards, in close boarding to roofs, boarded lining to wainscoting, sound boarding, &c. In heavier work it is found amongst a hundred other situations, in ordinary planking and strutting in sandy soils, between boards laid close to retain rubble filling, &c., and between planks that are intended to be caulked.

Cocked Joint.—This will be found explained under Caulked Joint.

Cogged Joint is the most usual term for the third variety of caulked joint, as previously described in this section.

Corked Joint.—This is one of the alternative names of caulked joint.

Cottered Joint is fully described under the Smith's trade in Section VII., and is much used by the carpenter in trusses for tightening up the tie-beam, in cofferdams for adjusting the cross-bracing, &c. Fig. 32 affords an illustration of its application.

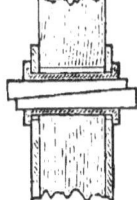

Cross Joint.—In planking foundations the planks are often laid in two thicknesses crossing each other and well spiked together,

Fig. 32.

the bottom thickness being laid diagonally across the masonry courses. In this and similar situations the planks are said to be laid cross joint.

Double Notched Joint is made by cutting a notch on each of the pieces of timber passing over one another, as in Fig. 33, so that they may fit down upon each other. The notches, however, must not be of sufficient depth to constitute halving.

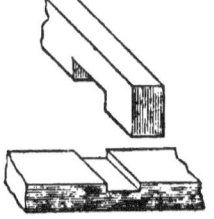

Dovetail Joint.—Dovetailing is shown in Fig. 40 and consists in a close-fitting union between a gradually expanding wedge-like pin or pins, or similar projections, cut

Fig. 33.

out of one edge or end of a piece, and a similar number of corresponding holes, notches, sockets, or grooves in another. It is more efficacious in stone than wood, owing to the shrinkage of timber in the direction of breadth and thickness loosening the joint and allowing the pin to draw. Moreover, when the pin expands from moisture there is a chance of its acting with the power of the wedge and splitting off the sides of the notch. Wood joists embedded in, that is, jointed to concrete, are more likely to remain firm therein if cut to a dovetail section. Halved joints are sometimes dovetailed, and in heavy work a mortise and tenon joint is strengthened by being dovetailed on one side, the other being tightly wedged up with a detached wedge, which is driven in to keep the dovetail in its seat.

Dovetailed and Halved Joint.—A description of halving in which a dovetailed notch is formed to prevent drawing, but which is untrustworthy for connecting ties. Timber shrinks about 2 per cent. in width and thickness, and consequently the dovetail in time withdraws from

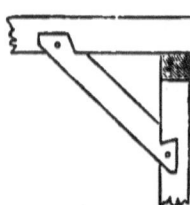

Fig. 34.

the cheeks of the notch and looseness ensues, which might work much mischief in some situations through the wedge-like shape and action of the dovetailed abutment of the tie. On the other hand, for struts, as in Fig. 34, the dovetailed notch is both suitable and picturesque, yet, at the same time, if the stress be considerable, it is not advisable to employ it for the reasons given under Strut Joint.

Feather Wedged Joint is the same as Fox Wedged Joint.

Fish or Fished Joint.—This is used for joining timbers in the direction of their length, and is formed by butting the square ends of two pieces together and placing short pieces of timber or iron, Fig. 35, usually of the same depth, and called fishing or fish plates or pieces, midway over the butt joint on opposite sides and bolting the whole firmly together. When of wood, the fish plates are sometimes tabled to the main pieces. If intended to resist a compressive stress, plates on all four sides are necessary, and so are straps or hoops if a transverse one has to be borne.

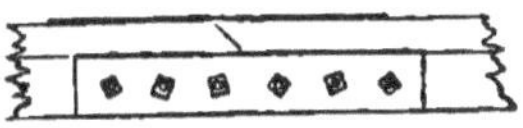

Fig. 35.

Fox Wedged or Foxtail Joint.—A mode of securing a tenon against drawing or the action of a tensile stress in a mortise that is not cut through, as shown in Fig. 36, by placing one or more thin wedges of straight-grained wood into saw cuts at the end of the tenon. On driving the tenon home, as soon as the wedges touch the bottom of the mortise, which is slightly dove-tailed or widened out, they split the tenon and compress it tightly home against the sides of the mortise. The more numerous and thinner the wedges are, the less chance is there of the tenon splitting up to its root.

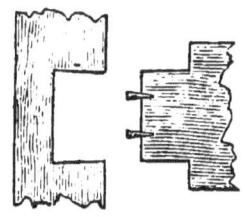

Fig. 36.

Framed Joint is one made with a mortise and tenon.

Grooved and Tongued Joint.—This is explained in Section VIII. Sheet piling is sometimes run with a groove 1½ in. wide and 1 in. deep to receive a tongue of hard wood the same width and double the depth, which is nailed to one of the grooves to stand out an inch, but if the piles are carefully fitted before driving, they will come into close contact when driven, and the swelling of the wood exposed to moisture will form a tight joint. There are, however, some cases when angular grooving or grooving and tonguing, as above explained, is highly desirable, and doubtless in some instances strong tongues between heavy submerged timbers assist caulking in keeping out water, and have proved more than once a source of consolation to the engineer. The floors of bridges formed out of strong timber planking should be tongued and grooved with tongues about 1¼ in. wide, and well caulked.

Half Lap Joint is another name for the common variety of halved joint.

Halved Joint is produced by halving or notching half through, and is used for connecting timbers that cross each other, and that either pass or not. Its characteristic is that a piece is cut out of each timber and down to half the thickness, and that when the upper timber is fitted down upon the lower, the top and bottom surfaces of both are flush. In only common or square halving, shown in Figs. 37 and 38, is an equal and corresponding piece taken out of each. The other varieties are known as bevelled halving, Fig. 39; dovetailed halving, Fig. 40; and dovetailed and housed halving, Fig. 41. In common halving, if the ends do not pass, they must be spiked or nailed to prevent a draw.

Fig. 37.

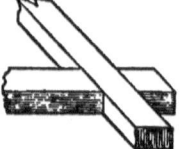

Fig. 38.

Fig. 39.

Halved and Bolted Joint.—If anything more than simple bolting is required, this or rebating and bolting is the best kind of joint for heavy timber work exposed to sea action, as it admits of tightening up, whereas the working loose of the tenon in a mortise joint cannot be remedied.

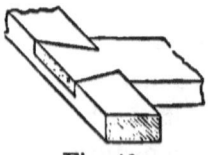

Fig. 40.

Heading Joint is one between the butting ends of two beams or timbers, and may be either a scarf or a simple butt, &c. In many curved ribs the heading joints of the logs interlace or are crossed so as to bond well throughout the layers. The same term has been given to the joint at the extremities of rafters when each pair is made to unite either by simple lapping, or with an angular half mortise and tenon angle joint, which sometimes occurs between principal rafters when framed together without king-post or ridge-piece.

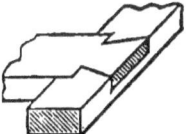

Fig. 41.

Housed Joint is similar to that of the mason, and consists in letting in the ends of timbers of any section or scantling into walls, cast or wrought iron boxes, heads, sockets, shoes, &c. Fig. 42 shows a mode of housing principal rafters into the head of a king-post whereby their heads butt against each other, in which position they are more likely to keep the king-post up than when the ends press against and compress its vertical fibres.

Fig. 42.

Housed and Dovetailed Joint is a variety of halving, or of notching when the pieces joined are not cut or sunk to the same depth. It is illustrated in Fig. 41, and is considered a stronger joint than that made by dovetailing only. The lip joint, Fig. 46, is another variety.

Inclined Joint.—This is an alternative term for oblique joint, under which name it is noticed.

Indented Joint is identical with a tabled or a tabled and indented joint.

Joggle Joint.—When a mortised piece is shouldered or expanded to receive the shoulders of the tenon, or when the cheeks of the mortise have jogs or notches to increase the abutting surface, such piece is said to

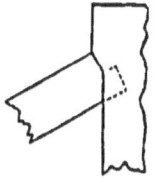

Fig. 43.

be joggled, and the joint into which it enters is denominated a joggle joint. This kind of joint is extensively used in carpentry for connecting the heads of diagonal or inclined pieces of framing with upright or horizontal ones, as shown in Figs. 43 and 44, and is used with or without iron fastenings. Another form of joggle joint is that produced between a short or stub tenon and its corresponding mortise, a dumpy tenon being sometimes known as a joggle. Some further remarks applicable under this heading will be found in the descriptions given of oblique and tabled joints. It is as well to observe, however, that a mortise is not essential to a joggle joint. Shouldering and notching appear also to be little else than varieties of joggling, for a shoulder may be formed by increasing or reducing the width or thickness of a piece, as instanced by the queen-posts of many trusses.

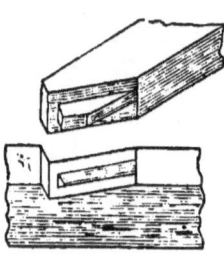

Fig. 44.

Key Joint.—This is made when slightly tapering wedges of hard wood called keys are employed to push home the different parts of joints and hold them up in their assigned places, in order to prevent too much strain falling upon the fastenings.

Keyed Joint is synonymous with one of the varieties of caulked joint, and has been already explained. It is likewise an alternative term for key joint.

Lap or Lapped Joint.—This is the simplest mode of lengthening timbers, and is made by laying one end over another and bolting through or strapping round with bent straps, having screw ends which pass through a bearing plate, or head or check plate to afford an un-

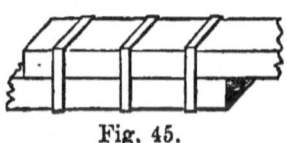

Fig. 45.

yielding surface for the nuts to screw up against. Fig. 45 shows the joint secured with hoops driven on whilst hot. Halving, however, is called lapping by some carpenters, and much that has been said under Halved Joint will here apply.

Lateral Joint is a longitudinal joint, and usually square edged, occurring between rows of planks in cofferdams, planking and strutting, &c. It also is found between the beams composing the ribs of timber bridges, and between the cushion and principal rafters of roofs of wide span. The carpenter uses boarding either rough or wrought on one or both sides, and joined lapping, or with edges square and shot only, or else ploughed and tongued, or rebated, or springed or sprung, otherwise bevelled. The latter is the best plan of jointing boarding for slating, for which and other similar work it is impolitic to use the best clean deal.

Lead or Leaded Joint.—A piece or plate of lead used formerly to be sometimes inserted between the abutting surfaces of timber joints to equalise if possible, and more fairly distribute, transmitted pressure, but there does not appear to be much ground for recommending the practice on that score, though with respect to the ravages occasioned by damp, the presence of lead in the joint has been known to exercise highly favourable functions. For example, in the Bridge of Bamberg, with a span of 208 ft. over the Regnitz, the timbers built into the abutments were thoroughly soaked in hot oil and then covered with sheet lead. Alfred Bartholomew, whose talent as a constructor is indisputable, and who well understood what he wrote about, though perhaps some of his ideas may appear to be slightly antiquated when viewed in the light of comparatively recent improvements and discoveries in building materials, appliances, and devices, observes with respect to bedding timber:—"If the plates lie upon stonework, there should always be sheets of lead placed between the stonework and all the timber, as stonework in general imbibes enough moisture to rot all wood in contact with it." The utility of the lead is no doubt connected in some way or other with the retention of the carbon in the timber, which is generally pretty safe from decay when dry and properly seasoned. Perfectly seasoned wood, however, of proper scantling for girders and the ribs of timber bridges is not easily procurable at the moment when wanted, and bedding timber upon lead prevents a fresh accession of moisture reaching the timber from the interior of the stone, whilst at the same time its non-porosity is equally efficacious in preserving it from the deleterious effects of sweat or a deposition of moisture attributable in certain conditions of the atmosphere to the conductivity and weak absorptive power of the stone. The plate of lead being cut to the exact size of the bearing end of a beam supported, for instance, by a stone corbel, would just raise the timber sufficiently above the top of the corbel to place it beyond the reach of any condensed vapour that might accumulate thereon, whilst all passage of wet from the wall to the beam would be peremptorily checked.

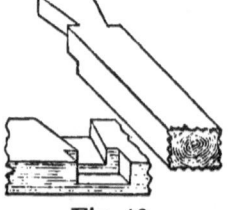

Fig. 46.

Lip or Lipped Joint.—This is furnished with a lip, as shown in Fig. 46, which helps to make a joint with stronger cross section than that already described as halved, housed, and dovetailed.

Longitudinal Joint is one in which the fibres of both timbers run parallel to the direction of the joint or seam, and is common in fishing and scarfing as performed to resist either a tensile, compressive, or bending stress. A lateral joint, as elsewhere observed, is likewise a longitudinal joint.

Mortise Joint is the junction of a tenon and mortise at any given angle. It is often called a mortise and tenon joint, under which term it is explained.

Mortise and Tenon Joint is formed by inserting a tenon into its corresponding mortise, so that the shoulders of the tenon bear upon the abutting cheeks of the mortise in vertical framing, whilst the side or edge of the tenon bears upon the cheeks or ends in horizontal framing. In either case the tenon should fill the mortise as completely as workmanship will permit; but when inserted horizontally into a level timber, it is strengthened, that is to say, relieved of much stress, by a shoulder or tusk, Fig. 47, the mortise being cut in the neutral part or the middle of the depth of the beam. In almost every instance the tenon is secured by a pin, key, or treenail (excepting when straps or bolts are required), whose diameter equals about one-fourth the thickness of the tenon, and whose distance from the shoulders is about one-third the tenon's length. A tenon in its simplest form is always a parallelopipedon in shape, and is usually made by sawing out an equal-sized notch on opposite sides of the end of a timber in the direction of its grain, as in Fig. 48. The plane of junction of the tenon with the rest of the piece to which it belongs is called its root, and the surfaces on each side of the root its shoulders. The mortise, which is accurately gauged on another timber to the exact size of the tenon it is intended to hold, is hollowed out with a chisel with its sides or checks parallel to the grain, Fig. 49, and perpendicular to the plane of the mouth of the mortise. In carpentry, in vertical framing, the tenon and mortise are each one-third the thickness of the timbers joined, these being in this

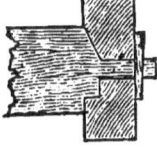

Fig. 47.

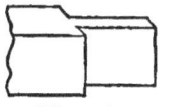

Fig. 48.

respect of equal dimensions; and the above description answers for the common case of joining at right angles,

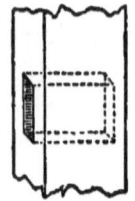

Fig. 49.

but sometimes the tenon and mortise have each one side notched or dovetailed, the remainder of the mortise not occupied by the tenon being filled up by a wedge. When the timbers are inclined to one another, the tenon is truncated, its shoulders bevelled, and the abutting cheeks of the mortise joggled as in Fig. 26, which shows the joint between a brace and post in a partition, or as in Fig. 44, which illustrates one of the numerous modes of uniting a principal rafter to a tie-beam. Double or parallel tenons are not much used in carpentry, the charge against them being that shrinkage and the imperfection of workmanship render it next to impossible to fit them so as to bear equally, whereby they not only thus weaken the framing, but produce an additional weakness by requiring two mortises side by side. Occasionally tenons are cut to fit mortises having a cross, L-shaped, or other form. The mortise and tenon is the fundamental joint of framing, though a frame or framing in carpentry is merely a collection of timbers united together by spikes, pins, or any means of connection whatever, and therefore not necessarily *framed* in the usually accepted meaning of the term, notwithstanding that it generally is. A truss is a frame having its parts so joined together by means of ties and struts that theoretically it cannot rack or change its form without rupture. When it consists wholly of timber its joints are almost always compound ones, consisting of a combination of the mortise and tenon with some sort of fastening, but when timber is in position moisture chiefly evaporates from the joints where the grain is cut across, so that a properly seasoned timber truss is far less likely to rack and fail than an unseasoned one.

Notch or Notched Joint is made when two pieces cross each other, and is used largely in letting down bridging and ceiling joists on binders, and purlins on principals, &c. A rectangular, or triangular, or other more complicated piece is cut out, leaving what is called a notch, which fits into another notch, or else down upon the uncut arris or side of another timber. Fig. 50 shows a common rafter in elevation and a section through the purlin, cleat, and principal. The purlin is notched and fitted down upon the uncut back of the principal, whilst the common rafter is notched and spiked upon the uncut arris of the purlin. Fig. 51 represents alternative modes of notching joists for spiking down to wall plates. There are other less simple varieties, but they more resemble forms of halving than of notching. Figs. 34 and 52 show struts notched into posts and beams. As a rule, in heavy work, struts should not be notched to horizontal timbers, but should bear upon cleats bolted so as to present an end grain for the seat of the strut, the two being connected by means of a rough tongue of boiler plate or stout dowel to prevent any lateral sliding at the joint.

Fig. 50.

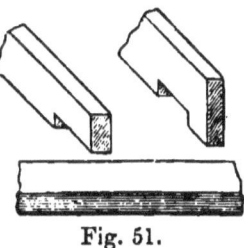

Fig. 51.

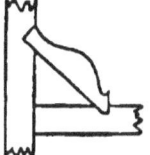

Fig. 52.

Notch and Bridle Joint is another name for a bridle joint.

Oblique Joint is formed at the junction of timbers or pieces which meet at an angle greater or less than a right angle. Oblique mortise joints are easily made by truncating the tenon and joggling the cheeks of the mortise, but in designing joints of this description, besides considering what may be the effects of the weather, lodgment of water, and the attacks of rot, it is well to remember that whereas the tensile strength of good fir is about one-sixth that of wrought iron of equal sectional area, its power to resist that description of shearing which forces the fibrous to slide over the vascular plates is only $\frac{1}{120}$, or thereabouts, of the shearing resistance offered by an equal-sized section of the metal in question. Hence mortise joints at the ends of really good timber exposed to oblique thrusts are generally considered weak, whilst iron straps, bolts, &c., present an admirable means of strengthening them. As a rule, the greater the obliquity of the joint the greater will be the shearing stress upon it.

Open Joint.—This occurs when the parts are not in contact. It is corrected in trusses, &c., by tightening up at the proper time with cotters or screws. If the joint happens to be exposed, and is caused to open and close by pressure and its removal, as is sometimes the case in roofs, timber bridges, &c., moisture will penetrate and cause decay, but a little management and skill displayed in weathering the joints and in obliterating all lodgments for water, so that none may flow into the joints as soon as opened, will go far to reduce the danger to a minimum. Lapping beams in exposed situations where practicable is better than scarfing, since there is less chance of the former kind of joint rotting from alternate dryness and moisture induced by the ingress of wet.

Piecing Joint is one effected by a scarf for the purpose of adding a piece having the same cross section to the length of a timber or plank, or of bringing up the width and thickness of any part to that of the rest by the insertion of a piece. It is made as described under Scarf Joint.

Pinned Joint is a joint strengthened or fastened with a pin or key. This kind of fastening is of frequent occurrence in carpentry to hold a tenon in place. In order to make the contact between the shoulders of the tenon and the cheeks of the mortise as close as possible, the hole for the pin is bored so as to draw the tenon, and the operation, which is called drawboring, is performed by boring the hole in the first instance through the cheeks only, then inserting the tenon and marking the position of the hole, after which the tenon is removed and the hole bored, not, however, in the place marked, but a little nearer the shoulders, and when it is again inserted, a slightly tapering conical piece of steel, called a drawbore pin, is driven through the holes, forcing the shoulders close home. On removing the steel pin, the proper one made out of iron or hard wood, sometimes wetted, is inserted in its place. Its use in carpentry should be quite of secondary consideration, stability being secured without such an auxiliary, and consequently the practice of drawboring can be of no advantage, but rather the reverse in well-conditioned framing. The diameter of a wood pin should not be less than one-fourth the thickness of the tenon, and its position should be at a spot distant from the shoulders equal to one-third the tenon's length. In shape, it is either round, round with a square head, or square throughout, which is considered the best form, as it brings more of the wood into resistance, and is not so likely to cause splitting if its sides are properly situated with respect to the stress. In order to render it as effective as possible against shearing, it should be split off from a straight-grained piece. When the tenon passes right through the mortise, as shown in Fig. 47, or as in the somewhat similar case of framing a trimmer into a trimming joist, the hole is bored so as just to clear the mortise, and the pin is driven against the outside of the joist. The term tree-nail (pronounced trunnel) is restricted by some to large-sized pins, varying in diameter from $\frac{3}{8}$ in. to $\frac{3}{4}$ in., and when of good English oak their resistance to shearing is about 2 tons per square inch of section.

Plain or Plane Joint is formed by the union or contact of two plane surfaces, as, for instance, when one piece of timber rests or beds upon, or abuts against, another, the area of junction being one and the same plane. It is characteristic of a fish joint. In common work, it is the usual mode of connecting joists to wall plates, rafters to ridge-pieces, &c., and usually a fastening of some kind is required to prevent sliding.

Post and Beam Joint occurs, as its name implies, between a post and beam, the latter being either inclined or horizontal, and either above or below the post. The points to be attended to are an equal and true bearing for the post, and the least possible diminution of strength in the abutment both of post and beam; and the choice of the best form of joint must in a great measure depend upon the nature of the load to be borne by the beam. In many situations the plan adopted in a common or proper door frame would no doubt suffice, viz., having a stub tenon at the foot and an ordinary one at the head of the post, whilst in others it would be preferable to cut a vertical mortise through the head and insert therein the tenoned end of the beam.

Pulley Mortise Joint is another term for a chase mortise joint.

Radiating Joint is made when the direction of two or more joints tends towards a centre or axis. For instance, the crossed joints of the back-pieces of centerings, and those of curved ribs put together in thicknesses, as well as the heading joints of circular curb-plates, belong to this denomination.

Rebated Joint.—This, like the similarly-named joint noticed in some of the other sections, occurs when a rebate or rectangular one-sided groove constitutes one of the meeting surfaces. Out of numerous instances of the adoption of the rebate in carpenter's work, may be mentioned its usefulness when gate posts are checked for heavy gates moving on central pivots to close snugly into them.

Rounded Joint.—A post or strut with rounded butment affords an illustration of this description of joint. It is calculated that with posts of equal length and scantling, the compressive resistance offered by square ends is three times that of rounded ones when left free to move.

Scarf or Scarfed Joint.—This is made by the operation of scarfing, which consists in uniting two pieces of timber longitudinally by similarly halving, toothing, or more or less intricately tabling and indenting both to the same templet, and then lapping their ends and tightening all up with folding or keyed wedges in a central mortise, (but not with such tightness as to instigate straining), and finally strapping or bolting with washers or iron plates. The simplest kind occurs in lengthening wall plates,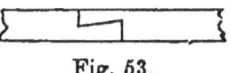

Fig. 53.

and is nothing more than halving the ends together with splayed surfaces without any fastening but spikes, Fig. 53, and out of many other complicated varieties may be mentioned those formed with a third piece of shorter length mutually connecting the longer ones, as shown in Fig. 54. The parts of each timber running with the grain, or nearly so, that come in contact in the joint are called scarfs, a term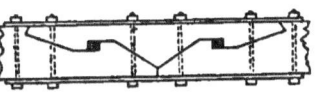

Fig. 54.

intimately associated with calf or kerf, signifying the wedge-shaped part formed by cutting the end slantingly. Scarfing usually runs in such a direction that when the lengthened beam is in position the planes of the tabling lie horizontally, or else inclined to its upper and lower edges, but when a transverse strain has to be encountered, it is better that these planes should run vertically between the sides of the beam, the joint being placed where the strain is not the greatest, or as far from the point of maximum strain as possible, or where if feasible the beam can be upheld by a post or suspending piece. Fish plates are employed to prevent the bolt heads and nuts crushing the timber, and not as in fishing, to form the bond. Sometimes, however, the ends of the fish plates are turned down and let into the wood to increase the tensile strength, in which case opposite ones ought not to enter or be too close to the same cross section, otherwise its effective area might be diminished to a risky extent. Straps and bolts are often used instead of bolts and plates. Scarfing, as a rule, but not universally, preserves the same width and depth of beam throughout, and it characteristically differs from fishing in the ends interlocking and mutually filling up their respective voids, instead of simply butting. It is used for piecing vertical timbers; but though some ways of making the joint will do equally well for compressive and tensile stresses, in most cases the nature of the chief force to be resisted must be considered before choosing the precise form. Oblique bearing surfaces are weak against compression, as they may slide and cause splitting with a wedge-like action, and, moreover, intricate forms are often weak and dangerous, for unless the projections all bear against their corresponding indents, which requires excellent workmanship to insure, the strength is reduced by each additional scarf. Roughly speaking, the length of the joint should average about four times the depth of beam when bolts and indents enter into it. Other varieties of scarfs are shown in Figs. 55, 58, 64, and 65.

Screw Shackle Joint.—This is similar to the joint of the same name described in the next section. It is used by the carpenter with tie-rods in strutting to floors, in composite timber and iron principals, centres, cofferdams, &c.

Serrated Joint is one with long indents or notches like the teeth of a saw, as represented in Fig. 55, and is sometimes used in scarfing and between the flitches of built beams, both straight and arched, to prevent the surfaces in contact from sliding. The different thicknesses, which have in some instances amounted to five or six or more layers of logs, partially interpenetrating each other throughout their length with serrated tabulations, are brought together by bolts and keys, but such an arrangement is safer for resisting a longitudinal than a cross stress, owing to the peril involved by the timber splitting across the serratures when strained by the latter description of force.

Fig. 55.

Ship Lapping Joint. — This is synonymous with halved joint.

Shouldered Joint occurs between two timbers when one is strengthened by being shouldered or thickened out to reinforce its abutting or resisting powers. Figs. 43 and 47 respectively show a mortise and tenon thus treated. The queen-post is often shouldered by cutting a notch to take the end of the straining beam, and the principal rafter is by some carpenters considered to be shouldered when a notch about 2 in. deep and 1 in. wide is taken out of it on both sides to give a bearing to the notched down purlin.

Socket Joint occurs in roofs, centres, cofferdams, and other heavy work, and is formed by inserting the tenons, ends, or other parts of two or more timbers or pieces exposed to a tensile or transverse strain into an iron receptacle shaped to receive them, and which is called a socket or socket piece. It is most frequently made of wrought iron, though it is sometimes cast, especially when employed as a "head." On the contrary, shoes which take the feet of timbers exposed to compres-

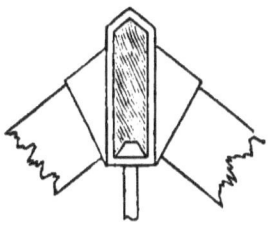

Fig. 56.

sive strain are usually of cast iron. One of its commonest
positions is at the apex of rafters, in which situation it is called a head, and is shown in Fig. 56, where a socket unites a king-rod, ridge-piece, and rafters, instead of framing the two latter into a timber post. Fig. 57 illustrates another one uniting a straining-beam, rafter, and queen-rod concentrated together, as usual in a truss, at the same point of support. Besides serving such purposes, a socket is useful in protecting the ends of timbers from damp or fire, and projections or cavities are easily cast or wrought upon it at any part to give support to cross timbers or other parts clustering around it.

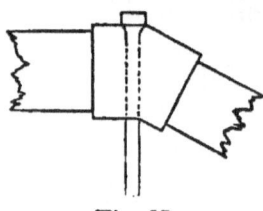

Fig. 57.

Spiked Joint.—This is one fastened with one or more spikes, which are large wrought nails varying from 4 in. to 14 in. in length. They are used for securing timbers to plates, shores or props to curbs, fender-pieces to walings, walings to piles, struts to posts, planks to beams, and, in fact, in most of the multitudinous descriptions of heavy carpenter's work.

Splayed Joint is a scarf in which the lapping ends are bevelled or cut slantingly, as in Fig. 58, so as to be oblique to the upper and under surfaces of the lengthened beam.

Fig. 58.

Splice Joint is another term for a scarf joint. Timber piles are sometimes lengthened by splices, often made with long bevels or splays strengthened with wrought iron bands and straps. Fish joints might perhaps be also allowed to enjoy the synonym of splice joints, since fish plates are known in some parts by the name of splices.

Square Joint is one in which the plane of the joint is parallel and perpendicular to the fibres of the two pieces respectively, as occurs when a post stands on a curb or sill. The feet of struts wherever possible should be cut square to the direction of their own fibres.

Stirrup Joint.—This is an abutting joint strengthened by a stirrup iron, which is a strap turned up from iron usually about $2\frac{1}{2}$ in. by $\frac{3}{8}$ in., and of the form shown in Fig. 59. The stirrup is secured with bolts and nuts, and keys or wedges, otherwise called cotters. The same shaped strap is passed sometimes round the feet of principal rafters, and bolted through the tie-beam to take the stress off the mortise in their effort to spread.

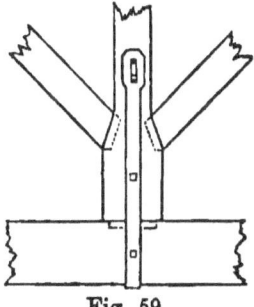

Fig. 59.

Strap or Strapped Joint.— This differs from a strap and bolt, or strap and cotter joint, inasmuch as the strap is continuous, and girts the work somewhat like a hoop does a cask, being either riveted or welded into a band, as in Fig. 45, or else having screwed ends which pass through a bearing or check plate, as with the kicking strap in Fig. 60, so as to be screwed down upon it with nuts. The strap is usually turned up from iron about $2\frac{1}{2}$ in. by $\frac{3}{8}$ in., and when of the welded description, requires to be put on hot, and the end of the timber over which it is slipped before being forced home to be slightly eased or tapered; besides which, dry weather should be selected for the operation. Either kind above mentioned may often be kept in place by coach screws.

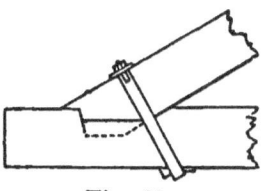

Fig. 60.

Lap joints, and those between the flitches or pieces of built beams, depend entirely upon the holding powers of the straps, when thus used, for their reliability, and consequently those of doubtful forging should be rejected without any scruples. When timber shrinks, in the course of time straps are liable to slip, and therefore in this respect, and especially when used with green timber, they are inferior to bolts.

Strap and Bolt Joint occurs when an abutting joint is secured with an iron strap and bolt, the strap consisting of a flat bent piece of wrought iron, from $1\frac{1}{4}$ in. to $2\frac{1}{2}$ in. wide, and about $\frac{1}{4}$ in. to $\frac{1}{2}$ in. thick. When used for connecting the foot of a rafter with a tie-beam, as in Fig. 61, it is indiscriminately called a heel strap or stirrup, and its two ends are usually swelled out into eyes to hold the screw bolt, which passes through the tie-beam near its neutral portion, and not so low down as the engraving represents. When the strap is attached to a suspending piece to support a tie-beam, it is termed a stirrup, the part which it embraces being all bolted through together, and secured with nuts and washers complete. This mode of fastening does not admit of adjustment. Variously shaped straps are employed in conjunction with screw bolts in framing together the component parts of trusses, partitions, domes, spirelets, bridges, and all sorts of heavy timber structures, but they are either turned-up pieces similar to the above, having equal and opposite ends, and corresponding holes formed therein for the bolts to fill, or else they may consist of loose duplicate pieces of identical form, having two or more tails or branches placed on opposite sides of the framing, so as to admit of being bolted together over the joint with the thickness of the framing between them.

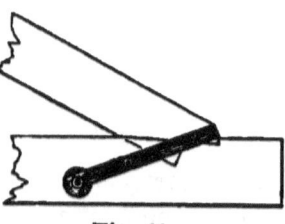

Fig. 61.

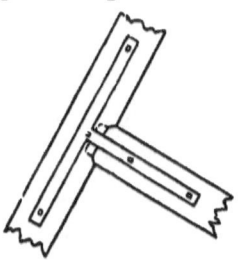

Fig. 62.

CARPENTERS' JOINTS. 107

In many cases, however, the duplicate strap is dispensed with, and washers used in lieu thereof. Settlement subjects this description of fastening, which has been previously noticed under Banded Joint, to cross strain. Fig. 62 shows one method of securing a strut to a rafter by such means. Straps and bolts, before using, should be preserved from oxidation, as remarked under Bolted Joint.

Strap and Cotter Joint is one strengthened with a strap, admitting of adjustment by means of wedges or keys, commonly called cotters, described under Cottered Joint, and which pass through slots in its turned-up ends, by which arrangement the effects of settlement or shrinkage may be partially counteracted. Sometimes only one cotter is thought sufficient to draw up the strap and the timber or element it grasps.

Strut Joint, as its name implies, is one between a strut and some other piece, and admits of different forms, the object being to obtain a uniform distribution of stress or an equality of bearing upon the whole of each surface in contact. If the beam is horizontal and fixed at one end only, a well-fitting joggle joint will answer; but if both ends are supported, the struts should, if possible, butt and mitre against a straining piece bolted to some part or other of the beam, as in Fig. 63. When the end of a strut is inserted so that it offers an angular rest, the consequences of splitting must be provided for, if the direction of the fibres appears to favour such a mishap.

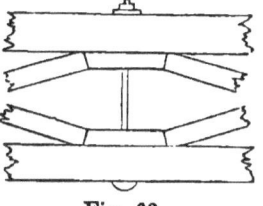

Fig. 63.

Swallow-tail Joint.—This is the Continental name for the dovetail, which, however, is occasionally so called in the United Kingdom. For instance, in dovetailing a joist to the depth of about ¾ in. down upon a wall plate, the resulting joint is often called a swallow-tail.

Tabled Joint.—The projecting surfaces of scarfs running in the direction of the grain are called tables, and the junction of two opposite and corresponding ones forms this joint.

Tabled, Indented, and Keyed Joint is a form of scarfing, of which common and useful varieties for resisting tension are illustrated in Figs. 64 and 65. On each bevelled or notched surface, as the case may be, an indent and a projection called a table are formed side by side, which respectively fit the corresponding and opposite table and indent in the other. Between the two tables, in a hole left for the purpose, folding and very slightly tapering wedges or keys of hard wood are driven in just sufficiently tight to bring the bearing surfaces well up together, which being done, the joint is secured with straps driven well home, or bolts and plates tightly screwed up, the latter being preferable to washers for preventing injury to the fibres.

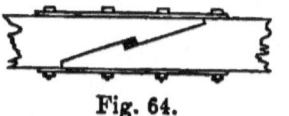

Fig. 64.

Fig. 65.

Tie Joint is one formed to resist a tensile stress, and amongst other positions, is found at the junction of a tie with timbers inclined to one another in order to retain them at a fixed inclination, or, in other words, to prevent them from spreading. It usually takes the form of a double notch slightly dovetailed (see Carpenter's Boast), as shown in Fig. 66, which must be spiked or pinned. Very frequently, when occurring between heavy timbers, iron bolts, sometimes jagged, or tough oak treenails, are useful to strengthen it against drawing. Wales or waling pieces are bolted to guide piles, and

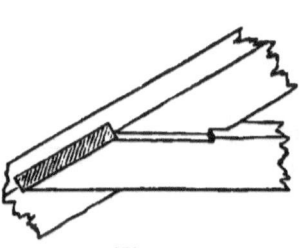

Fig. 66.

secured with nuts and large washers. Pile foundations to cast iron sewer pipes, &c., are tied together by the cap-sills, cross-heads, or bearers, which are either similarly bolted thereto with inch screw bolts, or bolted and checked. Iron straps bolted to timber, and having projecting corkings protected against oxidation, are built into masonry, and securely hold the two together.

Toothed Joint is a scarfed joint with serrated tables and indents, and has been already noticed under its alternative appellation of Serrated Joint.

Transverse Joint occurs when the fibres of one of the pieces joined are at right angles to those of the other. It may be either horizontal, as between a trimmer and trimming joist, &c., or perpendicular, as when posts, studs, puncheons, or quarters, &c., are connected with beams, plates, curbs, sills, &c.

Water-tight Joint.—In log huts the joints between the logs are made water-tight by filling in with clay, twigs, &c., and lining the inside with laths or boards; or simply packing the joints and plastering with clay will suffice. Joints between planks are rendered impervious to water by caulking them, as already explained under Caulked Joint.

Wedged Joint is one tightened up or secured by a wedge, which is a triangular prism having two planes meeting at a more or less acute angle, and forming what is generally known as the cutting edge. In carpentry, however, this edge is not required to be fine, but in joinery it is often all the better for being so. In driving wedges extreme caution is necessary, since the percussive action of the hammer so vibrates the particles of the body into which the wedge is being forced that every facility is afforded for its gradual penetration, which must end in splitting the body through the forcible separation of the opposite surfaces it slides against, unless stopped in time, or checked by a cramp or strap strong enough to turn the tables and bruise or cripple the fibres of the wedge, so as to destroy its elasticity and prevent it from swelling or recovering its size and shape after the removal of the confining ligature. In lengthening timbers by scarfing, folding wedges are much used, and they are as often as not called keys. Trimmers for carrying landings, and bearers for supporting winders, are firmly wedged to the wall. For tightening the cords of scaffolding, easing centerings, striking concrete platforms, &c., one or more wedges become necessary, as is likewise the case in many other departments of the carpenter's work, examples of which may be instanced in caulked joint, dovetail joint, fox wedged joint, &c.

SECTION VII.

SMITHS' JOINTS.

Abutting Joint.—In girder work, webs ought to be made to abut fairly against flanges. Where cast iron arched ribs are cast in segments, the abutting flanges of the joints should be truly planed, or chipped and filed and scraped to bear throughout the whole surface thus rendered fair and true, and which, of course, ought to be strictly at right angles to the curve of the rib or line of maximum pressure. In some instances the surfaces are payed over with asphalte previous to bolting together, and the lower arris chamfered off to obviate bearing on the edge. Sometimes the segments of cast iron ribs abut upon intervening plates or cross girders, which pass through the abutting joints of the parallel ribs lying in the same plane. In such instances, and in others where struts occur between the segments of the top boom or compressive flange of a girder, all the abutting surfaces of contact should be accurately and truly planed, or turned if of circular section. Cross plates, as well as the bed plates of piers and abutments, may have fillets cast on them to hold keys or wedges to secure by their careful adjustment an even bearing between the segments acting as voussoirs, but a correctly radiating and true and parallel facing to their abutting ends is a more essential expedient, and dispenses with such an artifice.

Adjustable Joint is a contrivance or union of parts

used for adjusting the lengths of ties and struts after the trusses they belong to have been fully weighted, &c. A cottered joint, explained further on, is one variety. Another kind consists in connecting screwed ends by means of

Fig. 67.

a nut, Fig. 67, having right and left hand threads cut in it at opposite ends, and sufficiently long to receive the extremities of the parts of the braces to be united, which have screw threads cut on them to fit the nut, the turning of which draws together or separates them as desired. Another great advantage of screwed ends is that they form excellent substitutes for welds, which unfortunately can never be relied upon. The kind of screw-shackle just described is called a box nut, but other forms are made use of. Where a compressive force only has to be encountered, the length of the strut may be adjusted by making it in

Fig. 68.

two parts, one of which is terminated by a long screw carrying a stout nut, as in Fig. 68, whilst the other part has one end bored out to a sufficient depth to receive the threaded end as far as the nut, whose position can be altered at will, allows it to enter. Another variety is described under Double Nut Joint.

Adjusting Joint.—This is obtained between the bearing plate and bed plate of a girder, or arched rib, or ponderous iron principal, by inserting between them in a cylindric seating a cast iron or steel roller at right angles to the longitudinal axis, as shown in Fig. 69, the arrangement for expansion being supposed to be at the other end.

Fig. 69.

The same motion can be equally well obtained by making the pin and either plate in one casting, as represented in Fig. 70. The joint thus formed provides for the tilting action set up at the bearings through the deflection of the girder, &c., under its load. The friction or bearing rollers in this figure give a sliding motion in addition to the springing motion arising from the hinging, and though it illustrates the application of the joint to a girder, the same principle is quite applicable to a roof truss. The bed plate under the rollers must be fixed with care to insure an even bearing, and firmly secured to the stone landings or other capping of the pier, or template on the wall, whilst the bearing plate that rests on the rollers requires to be ground or otherwise rendered smooth and truly horizontal.

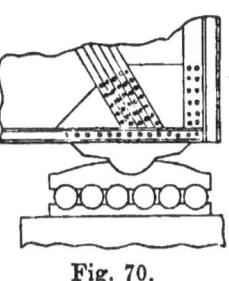

Fig. 70.

Anchor Joint.—More correctly, perhaps, Anchor and Collar Joint, which is the name given to a device forming the top hinge of a lock, dock, or other heavy gate. A strong piece called the anchor is firmly let into the stone work, and the collar of wrought iron is fixed to it, constituting a socket for the top of the heel post of the gate to work in. The joint occurring between any anchoring plate and its seat or bed is likewise an anchor joint, and may be as simple as that between a common plate or disc attached to a tie and the bulging wall it helps to sustain; or as grand as the junction between the anchorage plates of suspension bridges and the rock or immovable masonry on which they are seated. Iron skewbacks carrying spandril arches, and retaining walls steadied by anchoring plates, are kept in position by the reaction at the anchor joints at the further ends of their respective tie rods.

Angle Iron Joint.—This occurs when two lengths of

angle iron are united by a fishing piece, also of angle iron, as in Fig. 71, the whole being either bolted or riveted together. Similar joints are made in various ways with bolts and nuts; thus Fig. 72 shows an angle iron purlin joined to an angle iron rafter by means of an L cleat or short length of angle iron. Purlins in this or the reversed position, or rafters of angle iron, are often filled in with wood, unless the covering is to be of corrugated iron, or packings of wood are screwed or fastened to the iron work to fix boarding thereto. Slates, however, only require lead pegs or zinc or copper clips to unite them to iron laths, and boarding, therefore, in their case is not absolutely essential.

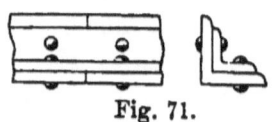

Fig. 71.

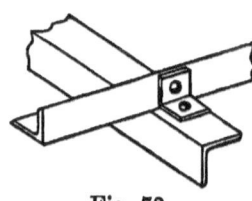

Fig. 72.

Angle Joint.—This is formed between plates when united by angle iron with rivets or screw bolts, according as the elements connected are wrought or cast. Square piers are thus formed by bolting together flat iron castings and building up the interior with brickwork in cement.

Arch Joint.—The cast iron segments of arched ribs in bridges form arch joints at their planes of junction, since the great additional strength gained by the curved conformation and radiating joints of the ribs is due to horizontal thrust, or in other words to their being *de facto* arches with long iron voussoirs, the necessity for an exact jointing of which has been already glanced at under Abutting Joint.

Bayonet Joint is useful for joining pipes or rods requiring occasional disconnection where longitudinal stress only has to be met. It consists of a socket or sleeve, Fig. 73, in which an L-shaped slot

Fig. 73.

is cut at the end of one of the pieces to take a stud attached to the other, which is slipped into it and turned round as far as the slot will permit. Circular gratings to drain traps are sometimes secured with this form of joint.

Bed or Bedding Joint.—Iron columns and stanchions may be bedded on stone or concrete by making the joint with asphalted felt, milled lead, iron cement, molten lead, sulphur, or Portland cement. With either of the two first a piece is merely inserted between the iron and base as a seating, but with the others the column, pilaster, or stanchion is first set up as nearly plumb as possible on its prepared base or bedstone, and correctly adjusted by means of iron wedges to the exact height and alignment, and then the bed joint is run with the cementing material, being further secured, if thought necessary, by bolts leaded to the base, as described under Bolted Joint, and by caulking where molten lead or rust cement is used. There appears to be no better plan than to set the column perfectly plumb with iron packings, and then to run the joint with neat Portland cement grout from the inside of the column, placing in the first instance a temporary curb round the base, so as to obtain a sufficient head of the semi-fluid to insure absolute solidity when set. Columns are sometimes cast with bed plates and bearing plates on the base and cap respectively, strengthened by vertical brackets termed feathers, stiffeners, or stiffening pieces, with bolt holes or caulkings, as the case may be, on the base plate for fixing to bed stones and holes in the bearing plate for bolting to iron overhead girders. The area of these expanded bearing surfaces should not much exceed three times that of the shaft, for if allowed to project too far danger will arise from cross strain in case of a slight yielding of the foundation, or the overhead girder deflecting. As a rule it is expedient to have the spreading base plate detached, as in Fig. 74, to obtain a perfect bearing and uniform distribution of load, rather than combined in the same casting with the column when large, and where this is the case the area of the base plate may be considerably greater than when it does not constitute a separate piece. Stanchions sometimes require specially cast spreading footings with lathe turned and truly faced surfaces of contact. Too much importance cannot be attached to the necessity of making the faced bearings and bed joints of pillars truly perpendicular to their axes, and fixing them sufficiently rigidly to prevent slipping in case of vibration or subsidence. Turned-up flanges or boxes, sometimes gracefully concealed with ornamental foliage, are cast on the cap plate when the columns support overhead girders of timber, which drop down upon snugs cast on the plate for a lateral tie. If there is a superimposed column, its base is cast with a flanged spigot for dropping into a socket, also flanged, formed by the prolongation of the cap of the lower column, the space between the bearing plate for the girder and the flanged and bolted bed joint of the columns being occupied by the floor. In many cases columns are bedded by direct bolting to massive cast iron base or foundation plates concealed below the ground-floor line. Iron bedded on iron requires true and parallel fitting and planing, or turning in a lathe, so that both head and foot, or top and bottom surfaces may be truly horizontal when fixed; and where a column fits into another, or only into its own base or capital, all the parts intended to be in contact should be turned and faced in a lathe, so that the proper touch may be preserved throughout. It is quite possible to chip and file, and finish off and bring the

Fig. 74.

surfaces of ironwork to as smooth a face as that of glass. When iron rests on stone, the latter must be sufficiently thick, dressed to a true surface, properly sunk for the plates of columns, or cut or bored for bolts, mortised for dowels, tenons, or caulkings, or sunk for the rivets of girders where the rivets are not countersunk. Cast iron girders carried by columns are sometimes forked out at the end to partially embrace the columns' head prolonged above the cap, a flanged necking or girder bed serving for their support, and a coupling link or shrunk-on collar forming the connection between two contiguous girders through the medium of burrs or snugs or stubs cast on them with that object. At other times the ends of the girders are flanged and bolted together over the cap. If a roof is to be upheld, the shoe or chair for the principal rafter is bolted to the girder near the column, which often acts as a downpipe. In bridge work a piece of thick sheet lead is sometimes inserted between the cast iron bed plates that are respectively bolted down to the granite bedding blocks capping the pier and the bottom booms of the girders. In other cases lead intervenes between the granite and iron, and in some instances the bed plates are planed to permit the girders to expand and contract, or they may be hinged, or else made to ride upon rollers according to the discretion of the constructor. Fig. 75 represents one method of bedding a superimposed column and overhead girder.

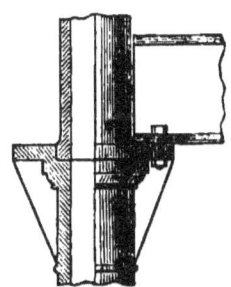

Fig. 75.

Bolted Joint.—In connecting the parts of cast iron structures, bolts and nuts are indispensable, since riveting is out of the question, on account of the hammering required for the operation and the contraction of the rivets in cooling being too much for a metal of such brittleness and inflexibility to stand. The efficient manufacture of bolts, which is at all times of consequence, is, however, doubly important in wrought iron structures, where they are little used excepting when large enough to insure soundness. As a rule, bolts should be made from one piece, turned to exactly fit the holes drilled to receive them, and case-hardened, together with their nuts and washers, for superior work. To meet severe tensile strain, bolts are more equal and reliable with nuts at each end instead of the usual head and nut, and where very violent strains can be foreseen they should be of steel. It is essential that in every case the abstraction of strength caused by the bolt hole is compensated for by an equivalent increase of metal where necessary. In cast iron elements this is often accomplished by thickening the metal round about and a little beyond the region of the bolt hole. To the same end cast iron beams, &c., have flanches, and other pieces, lugs, or ears cast on them with the usual bolt holes, whilst where two parts cannot be directly joined together, intermediate pieces, called brackets or dogs, are cast with proper holes with the object of being bolted to both, and all additional holes that may be required in any of the parts must be drilled; but as the perforations weaken the metal, and as no additional strength can be imparted to the casting by increasing its mass, the utmost caution may, perhaps, be necessary before determining the points for the extra holes. In cast iron bridges the several segments are bolted together through plates set between the planed surfaces of contact to enable the cross-bracing, or stays, or cross girders to be riveted or otherwise attached to the main ribs. The application of bolts with round, square, or countersunk heads to fasten the parts of iron work used in building is a wide one. Cast iron chairs or shoes securing the feet of iron rafters are connected by bolts, as are likewise similar castings at times fixed to rafters for taking the purlins. Tie-rods, and king and queen rods are also bolted to other parts of the trusses which they brace, having sometimes jaws at their extremities through which pass the bolts to hold the parts they seize. In these situations the effective diameter of the bolts must be proportioned to resist transverse strain. Cast iron distance pieces are bolted between double bars of different section to obtain increased strength and stiffness when the parts are long, or timber is flitched between their whole length instead. Side and central cast iron standards for supporting ventilators, raised skylights, &c., are bolted to the rafters. Corrugated iron is fastened to iron purlins with small galvanised bolts, and different lengths of eaves troughing, and ornamental guttering are rendered with red lead or putty at their lapping ends, after which the joint is secured with nuts and gutter bolts about ¼ in. in thickness. It will be observed that in various parts of smiths' work bolts are not screwed up tight, but allow a little play or springing motion, in which case the joints of which they form a part are movable and not fixed. If bolts and nuts are to be protected against rust by Barff's process the male and female screws must be cut larger by an estimated amount of increase.

Brace Joint is met with notably at the junction of cross-bracing with screw piles, columns, &c., in wind-bracing to roofs, &c., and may be either riveted, keyed, or bolted according to circumstances. The bracing between the booms of main girders, which are usually of wrought iron, is generally attached by riveting, though pinned joints are not uncommon.

Brazed Joint.—This will be found described under the Coppersmiths' Joints.

Broken Joint.—In riveted and bolted structures broken joints are supposed to impart additional strength.

In building up girders, &c., a broken joint consists in allowing no two plates or angles to approach coincidence in any cross section. When the butts of all the members occur in the same section the butt is called a full butt, but there is a difference of opinion as to the propriety of such a joint, yet its advantage is obvious in enabling heavy girders to be completed in detached portions. Though broken joints give, as a rule, an augmentation of rigidity, it does not necessarily follow that, because the joint is the weakest part, the weakness is diminished by transferring part of the joint to another cross section taken at random. As a curious instance of broken joints, it may be mentioned that the tie-bolts of cofferdams, instead of traversing the whole thickness, have been made to break joint by being bolted to an intermediate row of piling in order to lessen the orifices by which water invariably and annoyingly finds its way to the interior.

Butt Joint.—This term is applied in riveted iron work when the ends of plates, &c., are placed against each other and fished with a cover either on one or both sides to keep them in position. If the joint is in tension and the cover plates are on both sides the rivets will be in double shear, and they need not be so numerous nor the plates so long as when only a single plate is used. In girder work the rivets of cover plates of joints in tension may be advantageously reduced in number towards the ends and centres of the covers. The weakest cross section of any prism is of course that which is most perforated, and in case of a main or cover plate tearing away along this section, if there be on each side of it other rivets presenting sufficient shearing resistance, and so situated by a gradual reduction in the number per row as not to equally weaken any other section, there will not only be a considerable margin of strength left, but the weakest part of the joint will not be through the weakest section of either main or cover plate.

Much skill and nice calculation are, however, required in spacing and distributing the rivets in a butt joint in tension, whereas superior workmanship is the chief desideratum when compressive stress has to be borne, in order that the pressure may pass evenly along from one main plate to the other without the slightest break, either at the rivet holes or ends. The latter of these must be planed square to the direction of the pressure and fair and true to gauge, and not too short or they will be open. As this Utopian perfection, however, is well-nigh unattainable, some engineers elect to design joints in compression as if they were going, in fact, to be racked by tension. Further notice is taken of this joint under the heads of Caulked, and Double Riveted, and Single Riveted, whilst the term full butts has been explained under Broken Joint. Fig. 76 represents a butt angle joint, secured by a bent bolt riveted to one plate, and screwed up with a nut to the other. A butt joint in tubing requires hard solder (explained under Brazed Joint) to be durable.

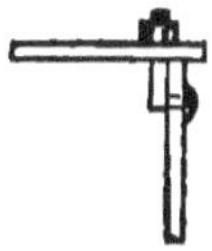

Fig. 76.

Cast Iron Joint, as its name implies, occurs between cast iron elements, but the term, though useful enough, is not often met with. As a rule, the surfaces forming it wherever transmitted pressure passes through them should be faced truly parallel by planing or chipping and filing, to insure equal bearing throughout. Broken castings can in many instances be repaired by soldering, as elsewhere explained.

Caulked Joint is water-tight, and occurs in piping, wrought and cast ironwork, &c., and is made by stuffing the cavity in the socket, or between plates, castings, &c., with oakum, brown paper, and white or red lead, india-rubber, wood wedges or slips, iron or rust cement, or gaskets of spun yarn, either white or tarred. One or other of these materials is vigorously driven into the cavity or crevice with a hammer and thick steel chisel, of different

G

shapes and sizes, called a caulking iron or tool, excepting, however, the wood and india-rubber, which do not require the exercise of much force. The joints of box girders are sometimes caulked to render them air-tight, to preserve the interior from corrosion; and in other riveted work, fished joints are not unfrequently caulked by punching down with a caulking tool the butts or edges of the plates drawn apart by the process of riveting until they are brought close. In a similar way, the heads or points of rivets and the junctions of plates in water-tight work are caulked down if not left sufficiently close. When the pitch is very wide, the plates will spring under the operation without great care. This kind of joint is otherwise written calked or cauked. There is another variety, however, which, though identically named, is essentially different in purpose and principle. It is formed when a caulking enters into the composition of the joint. This term caulking, calking, cauking, or corking, is given to an iron fang or projection, or flattened end turned up or down, or both (common in chimney and tie-bars), for building into masonry, as well as to a tenon or projecting stub cast on columns, pilasters, standards, girders, &c., for letting into bearing or bed plates, bedstones or blocks. Hence the resulting joint is obviously either calked, caulked, cauked, or corked.

Cement Joint.—Red and white lead in equal parts mixed with linseed oil forms a good cement for hot-water pipes. Under Rust Joint will be found described another variety. Eaves gutters, &c., are jointed to remain water-tight, as noticed under Bolted Joint.

Chain Riveted Joint.—A double, or triple, &c., riveted joint formed either by lapping the main plates, or by means of cover or fish plates, the characteristic of the joint consisting in the rivets being similarly

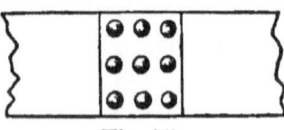

Fig. 77.

and equally distributed, and standing regularly behind each other, along equidistant lines parallel to the direction of the stress. Fig. 77 shows a triple chain riveted lap joint, in which, however, one of the cross lines is erroneously drawn full instead of dotted.

Chamfered Joint consists in chamfering off the arris of bed plates, &c., to present to the superincumbent pressure as obtuse an edge as possible around the bed without appreciably diminishing its bearing surface.

Chipped and Filed Joint.—This is effected by knocking off the hardened scale or protuberances on the surfaces of castings, &c., by means of a hammer and cold chisel, called a chipping chisel, having a slightly convex face, and a cutting edge of about 80°. The chipping is followed by filing down, which finishes off the surface to a fair face to obtain a true and equal surface and close joint. The planing machine has to a great extent superseded this method, but it cannot always produce so true a surface as filing and scraping, nor can it be said to be a mightier agent than chipping, for no mass of iron, either wrought or cast, and however hard or tough, can withstand the onslaughts of the chipping chisel. The machine, however, like most others, has its own distinguishing merit of facilitating work and economising labour.

Circular Joint is formed when a convex foot or bearing plate is bedded upon a concave bed plate, or the reverse of this will hold. An iron column under such conditions if not fixed but allowed radial motion is said to lose two-thirds of the strength it would otherwise possess if bedded with a flat bearing. So far, however, as its application to provision for expansion goes in combination with a small allowance for deflection, or springing motion, a circular joint answers excellently well, the lower plate being furnished with a roller bearing, whilst the upper is bolted either to the girder end or rafter foot.

Clamped Joint.—One held fast by clamps consisting of opposite pieces riveted to a plate, &c., in order to seize between them the part to be confined or joined.

Clip Joint.—This occurs at the angles of wrought iron chain bond and hoop-iron bond when the ends are turned up and down to clip one another. It is also formed by a contrivance called a clip, consisting of an embracing strap or else of a pair of straight, or curved, or bent arms or edges. Bracing-bars are joined to wrought iron piles or columns by means of clips which surround them and project as ears or lugs with an interspace between for the insertion of the bars, the whole being bolted through together. The bars are either of angle or T-iron, &c., or else round with flat ends and eyes. When the columns or piles are cast the ears or lugs for the cross bracing are cast on them at intervals in the height. The diagonal bracing bars of submerged piling and overhead columns, which act as ties or struts, must always be made suitable for resisting compression and tension respectively as well as their own proper stresses, though in order to impart the requisite rigidity and stiffness to the bracing the ties should be adjusted to throw the struts into compression. It is on account of the variable and uncertain nature of the possible disrupting forces that provision should be made for the braces to act indifferently as ties or struts as occasion may require. Cast iron sheet piles, $\frac{1}{2}$ in. thick and nearly 2 feet wide, have been successfully secured and jointed by the agency of clips.

Close Joint.—This will be found noticed under Tight Joint.

Collar Joint.—Long rings or detached sockets with internal threads connecting lengths of iron barrel, tie-rods, &c., are termed collars, and form collar joints. Lengths of copper or wrought iron chain-bar, 2 in. or 3 in. wide by $\frac{3}{4}$ in. thick, built into stone architraves, or round towers, domes, &c., and secured to each stone by stubs leaded in, are sometimes joined to each other by means of collars of the same metal, likewise leaded into the stone to receive the turned-down ends of the bars, which, however, should always be in as long lengths as possible, the joints invariably being over the columns in the case of entablatures. When of large size, collars used in piping are not threaded, but fixed by caulking so as to cover the butting edges of two spigot ends, the collar being easily slipped over the pipe and back again over the joint after a piece has been cut out to make room for a T-piece. A very efficient kind of collar has been used for uniting large waterpipes or mains. The pipes are cast with both ends beaded and no socket. These are butted together and each contiguous pair of beads surrounded with a well-tallowed band of felt about 6 in. in width, and this is again encompassed with a cast iron collar made in 3 or 4 flanged segments, which are bolted together, closely gripping the beads and forming a tolerably inexpensive, sound, and somewhat flexible water-tight joint.

Corked Joint.—This has been already explained under its alternative appellation of Caulked Joint.

Cottered Joint is formed by means of two folding iron wedges called cotters, which have a "draw," and pass through slots made in the plates or flattened-out ends of the rods or parts they unite. In order that the cotters may glide smoothly on their seats, wrought iron shields called gibs, having turned-up ends to clutch the ends of the slots, are inserted in them, as shown in Fig. 32, which represents the gibs and cotters of a strap upholding a tie-beam to a king-post. Sometimes a single cotter is used without gibs. The angle of obliquity of the cotters, in order to be safe against slipping, must be very little, and should not exceed 4°, which is the angle of repose for greased surfaces of iron upon iron. Cottered joints are used for adjusting the length of tension rods, diagonal ties, king-bolts, &c., the ends of these pieces being either forked to seize exteriorly the parts to which they are fastened, or else flattened out and inserted between their sides, the cotters and gibs passing through all. In the attachment of tie-rods, the cotters often go directly through a slot in the cast iron shoe to which the rafter is bolted. The shoe may be either on the cap of an iron column, or on the top of a wall, or it may be bolted to the inside face by means of a horizontal bolt secured to a plate outside and passing right through the wall. If the shoe rests on the top of the wall, it is secured to a stone block with lewis or dovetail bolts leaded thereto, or else vertical bolts with corkings are built in the brickwork that traverse holes drilled in the stone as well as others in the shoe, which is then firmly bolted down. If provision has to be made for expansion, it is done as described under Expansion Joint. Where no adjustment is required, the end of a stay bar may have a loop so made as to pass over a socket, through a slot in which a cotter is driven to confine it.

Coupling Joint is formed by means of a coupling, which is a device for connecting parts together, and almost as diversified as joint itself. Amongst builders it usually takes the form of a hinge, union, screw shackle, fish-plate, detached socket, collar, or sleeve, while pump-fitters use a key joint for coupling well-rods. Tension rods in roofs and horizontal bracing in bridges are usually regulated by screw shackles. Parts of wrought iron piles or pillars, when made in more than one length, are connected by a coupling-piece consisting of a short wrought iron or steel cylinder, which is slipped over the butting ends, as shown in Fig. 78, and bolted to both in cases where welding is not feasible, but great care is needful in the case of piles that the bolts and bolt holes are sufficiently

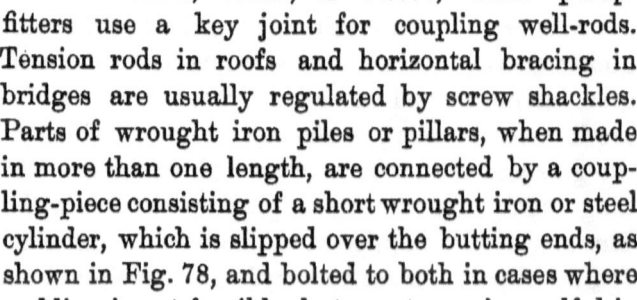

Fig. 78.

firm and reliable to withstand the great twisting strain that will be brought upon them in screwing in.

Cover Joint is a butt joint covered by a plate on one or both sides.

Cramp or Cramped Joint.—This is an alternative term for clamped joint.

Diagonal Joint occurs in riveted cylindrical iron structures, and is made by riveting the plates together when spirally arranged round the axis with which the joint forms an angle of 45°. It is shown in Fig. 79, and in the case of a boiler is one-fourth stronger than a longitudinal joint. The same term

Fig. 79.

for want of a better is applicable to the connections of the diagonal bars of a girder with the flanges or booms. The joint may be made either with one large pin or two or more rivets, the latter being safest, but bolts or pins are compulsory with cast iron elements. Ornamental castings are employed to conceal the diagonal joints of roof trusses.

Double Nut Joint is one formed with a nut and a back nut. Thus a suspending rod passing through a flattened tie-rod is often secured by a nut on both sides for purposes of adjustment.

Double Riveted Joint.—A joint in ironwork formed by a double row of rivets through both plates either arranged as chain-riveting, Fig. 80, or as zigzag-riveting, Fig. 94. The plates may either overlap or be joined with cover strips or plates—that is, may either form a lap or butt joint—in which latter case there will be four rows of rivets; if but one cover be used, it must equal the main plates in thickness; but if two, then each must not be less than half the thickness of a main plate. The

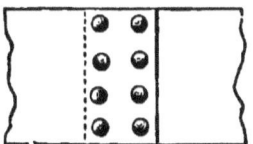

Fig. 80.

following gives a tolerably exact idea of what is necessary

to insure equal strength against the crushing and tearing of the plate on the one hand, and the shearing of the rivet on the other; but in practice engineers have no absolutely uniform rule. In a double-riveted lap-joint the diameter of the rivet equals two thicknesses of plate, the pitch (or distance from centre to centre of rivets in the same row) is from $4\frac{1}{2}$ to $5\frac{1}{2}$ diameters, and the distance between the pitch lines (or lines through the centres of the rivets of each file or row) should be in zigzag-riveting two-thirds of the pitch, and in chain-riveting $2\frac{1}{2}$ diameters, the total lap for the two kinds being $5\frac{1}{2}$ and from 6 to $6\frac{1}{2}$ diameters respectively. In a double-riveted butt-joint the diameter of the rivet equals $1\frac{1}{8}$ the thickness of the plate, the pitch is $5\frac{1}{4}$ diameters, and the distance between the pitch lines in chain-riveting is 2 diameters, and in zigzag-riveting 3 diameters. Hence the width of the respective cover-plates is 11 and 13 diameters.

Dovetail Joint.—This is found occasionally in smiths' work, and an instance of its application occurs at the junctions of the ends of window bars in some varieties of wrought iron frames.

Dowelled Joint.—Wrought iron girders carrying floors are, amongst other contrivances, jointed to walls by means of a wrought iron dowel about 1 in. square and 3 in. or 4 in. long, leaded to the seating, a slot being cut in the girder flange to receive the same and allow a little longitudinal play. Rafter-chairs are likewise similarly secured.

Drilled Joint.—A riveted joint in which the rivet-holes are drilled and not punched. Before drilling, however, all the parts of the girder should be temporarily put and held together with a sufficiency of bolts, and then the holes should be drilled to gauge through all the thicknesses, at the same time care being taken that all burrs left by the drill on each plate or piece are removed by slight countersinking before riveting. The holes thus made may be closer than when punched. A drilled joint costs more, and is supposed in some quarters to be weaker than one that is punched, though the drill being a cutting tool does not treat the metal so roughly as the punching machine. It leaves, however, the edges of the rivet holes sharper and firmer than when punched, which perhaps enables them to shear the rivets with greater ease. There is much to be said both for and against either practice. As regards their relative merits, it is sufficient to say that the conflict of opinion points to no surplusage of advantage on either side, and a safe and prudent course therefore is to drill the holes at the joints and where there are several thicknesses of plate and to punch elsewhere. In cast iron work, as already observed, holes if not cast must be drilled.

Elbow Joint.—This is formed by means of an elbow piece, one form of which is shown in Fig. 81.

Expansion Joint. — As stated in the Plumbers' Section, this form of joint is necessary to prevent accident and injury arising from the natural elongation and shortening of certain parts of structures exposed to changes of temperature. A difference of 60° Fahr. in its temperature alters the length of 200 ft. of iron by the space of 1 in., and in exposed situations a minimum provision should in consequence be made at the rate of about 2½ in. for every such length to meet its fluctuations. Some engineers, however, consider an allowance of ½ in. per 100 feet sufficient. The range of temperature in this country is about 83° Fahr., and a variation of 1° is theoretically equivalent to a change of load amounting to 1¼ cwt. per square inch of section, but the temperature of ironwork does not vary by a similar gradation to that of the atmosphere and therefore its effect upon a girder, &c., nduces less additional stress than would at first sight

Fig. 81.

appear probable. As regards roof principals, in order to allow a little room for sliding crosswise on the wall, the bolt holes are elongated in the cast iron rafter-chairs on one side of a roof, or else the chairs are bedded on a plate of 9 lb. lead. An iron dowel, about 1 in. square, is, under these circumstances, sometimes leaded to the stone template, and entering a slot in the chair, duly limits its movements. If a long girder is bolted to a pier in its centre both ends are left free, but generally one end is fixed, the other being supported so as to allow slight motion. This is managed by merely planing the cast iron bed-plates respectively bolted to the girder and abutment, and lubricating the surfaces, or else by inserting between the plates from six to eight cast iron or steel bearing or expansion rollers fitted together in a wrought iron roller-frame superimposed upon a cast roller-bed, in the manner indicated in Fig. 70. Arched ribs and wide roofs, either curved or span, likewise require a similar bearing on one side for sliding motion, and usually it is necessary in works of magnitude that over the rollers there should be an adjusting joint to insure equal bearing upon them, together with due freedom to the structure as regards springing motion. In the Clifton Suspension Bridge there are jointed ends or flaps 8 ft. long at both extremities of the roadway to give vertical and longitudinal freedom of motion to meet the effects of contraction and expansion as well as those arising from high winds and heavy loads. Fig. 87 represents a more or less satisfactory contrivance for meeting the requirements of variable length in a bridge rib. Cast iron becomes insecure at 32° Fahr. in the face of jar, vibration, or tensile strain.

Eye Joint.—When a rod or bar is swelled out at the end and formed into an eye or hole for a bolt or pin, the joint so effected is an eye joint. In simple trusses bolts or small pins are commonly used for connecting the parts of composite ties, which are accordingly provided with holes or eyes at their extremities, and the same sort of connection in heavier work is noticed under Pin Joint.

Faced Joint is formed between cast or wrought iron elements or parts, such as the flanges of columns, pilasters, cylinders, tubes, &c., or between bearing, base, bed, or cap plates, flanges, &c., previous to bolting down, riveting, or bedding. It is made by levelling and smoothing the surfaces of contact by chipping and filing, or by the use of the planing machine, with the object of securing the benefits of a full and continuous bearing and a consequent equal or symmetrical distribution of transmitted pressure. In some cases wrought iron girders are riveted in complete divisions or sections, which, being faced at their ends to the proper obliquity, are built up in much the same manner as the segments of cast iron ribs. The caps of piles are faced in the lathe to afford a true bearing.

Faucet Joint. — This is synonymous with Socket Joint.

Ferrule Joint occurs when a ferrule or small hollow metal cylinder, prism, or cone enters into the joint, or, in other words, forms one of the meeting surfaces. When slipped over a bolt and used in conjunction with it to keep at a fixed distance apart the surfaces bolted to one another, it may be flanged. In iron fencing, a ferrule joint is made when a ferrule similar to that represented in Fig. 82 is passed through a hole in an iron standard to hold the ends of horizontal round or square bars, which are inserted and riveted therein. It is also produced between a main and service pipe, when tapping the former, by

Fig. 82.

driving or screwing a piece called a ferrule, usually of brass and either straight or slightly cranked, or of elbow form, into a hole pierced with a driving punch or drill in the main to receive it. A driving ferrule is formed with a slightly conical end and a lug or shoulder for hammering on. In punching a hole, its site is first chalked on the main, and previous to punching out, its outline is closely dotted round with the punch so that there may be no risk of exceeding the required size. The hole thus pierced being jagged, it is rimered out to a true circle with a six or eight-sided rimer. The best plan, however, is to drill the hole with a drill clutch and ratchet, or other brace. When a screw ferrule is used, the hole made either way must be tapped with a full and perfect thread to match that on the ferrule.

Fish or Fished Joint is formed when two plates meeting at a butt are fished with a cover plate on one or both sides, the whole being riveted or bolted through together.

Flange Joint.—This is made by uniting together projecting rims called flanges, as shown in Fig. 83, with rivets or bolts and nuts, the former of which are not used with

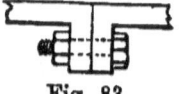

Fig. 83.

cast iron. The abutting flanges of cast iron ribs and the external or internal flanges of large columns require to be truly faced before bolting. Iron columns supporting overhead iron or timber girders in warehouses, &c., are provided with large flanges in the form of bearing plates or expanded caps, strengthened with stiffeners, and these require careful facing before attaching the ironwork; but where flanged cylinders are superimposed upon one another from the foundation upwards, and filled in solid with concrete or a core of masonry, which is carried up an inch or so above the topmost cylinder for bedding the foundation slabs of the superstructure upon, there need be no nice adjustment of the flanges. Not less than four and not more than eight bolts are, as a rule, allowed to each flange. Lengths of cast iron pillars and uprights are generally united with bolts passing through flanged rims. The segments of built columns require bolting together, so that neither bending, nor buckling, nor sliding at the vertical flanges may occur. Cast iron cylinders, bolted together by means of inside horizontal flanges and filled with concrete, have split all round their circumference after severe cold, which has been accounted for from the sections through the flanges contracting more than where the metal is thinner, and the concrete not yielding proportionately to the contraction of the flanges. When necessary in piping and elsewhere, the joint is rendered tight by inserting between the flanges india-rubber washers, rings, or strips, or a composition called red cement, consisting of two parts white lead mixed and one part red lead added dry, the whole being worked up to the consistency of putty. Tar and gasket, with or without iron washers, or iron-wire rings let into opposite annular grooves in the flanges, or rust cement, described under Rust Joint, are likewise used according to circumstances. A ring of metal wrapped round with woollen cloth, then dipped in tar and inserted between the flanges, is well suited for pumps when large. Thin wood fillets or slips are similarly useful and efficacious as packing between the flanges of the elements of cast iron cisterns. In ironwork, angle iron sometimes forms a convenient flange when one has to be attached after rolling or casting. Flanges are generally used for pipes set vertically, and are handy when joints have to be loosened, but they are not adapted for some situations exposed to variable temperature.

Folded Angle Joint.—One variety is shown in Fig. 84, and is used with solder. Fig. 85 shows another form common in simple kinds of ironwork. The edge of one plate is bent to an angle to fold over the other, and the two are connected with small screw bolts.

Fig. 84.

Grooved Joint.—This is formed by means of a groove, as, for instance, when piles or cylinders for submerging are cast with vertical lateral indented grooves, into which plates of timber or iron are slid to form a temporary or permanent barrier to the passage of water.

Fig. 85.

Grouped Riveted Joint occurs when two or more layers of main plates are used in one thickness, as in

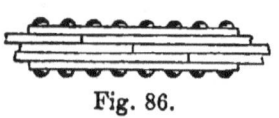

Fig. 86.

Fig. 86, with the view of collecting as many joints as possible under one pair of cover plates. Each butt joint must be so arranged as to lie as far as practicable from the others, and to be at the same time well within the cover plates.

Hinge Joint is made when the connected parts are fastened with a pin or bolt constituting an axis, and allow-

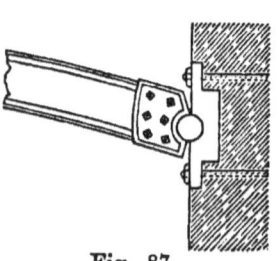

Fig. 87.

ing play or motion to ensue within definite limits around it. Out of many forms may be mentioned that which occurs when hinged shoes consisting of steel, cast iron, or gun-metal bearing plates, with an interposed pin, are fixed over the points of support to enable girders to ride and transmit their resultant pressure centrally down columns or equably over expansion or bearing rollers, as in Fig. 70. In the case of arched ribs, the bed plates of the hinged joints are built vertically into the abutments, as shown in Fig. 87, the hinge forming a provision against small longitudinal changes.

Hook Joint.—From numerous applications of the hook in jointing may be selected that occurring in fireproof floors when interties are secured by bending their ends into the form of a hook so as to pass up by the side of the web, and then over the top of the girder so as to give it lateral strength, whilst other cross bars are hooked on to these at right angles to them so as to form a network of rodding to receive concrete or plaster.

Horizontal Joint.—This necessarily occurs at, and is identical with, a level joint in cisterns, bridge piers, and other cast iron structures.

India-rubber Joint is made by inserting rings of vulcanised india-rubber in pipe sockets to render them water-tight. The plan does not appear to be popular for large pipes, but in the case of low-pressure water pipes for heating conservatories, &c., a short tube or long cylindrical ring of this material is stretched over the spigot, which is then pushed into the faucet, and if thick enough to well fill the cavity, will make a perfectly tight joint as soon as the water circulates. Joints, however, of this nature must not be made too near the heating apparatus, pure caoutchouc fusing at about 360° Fahr. Flat rubber rings make a good joint also between flanges.

Iron Cement Joint is one stopped with iron cement. This composition is prepared as follows:—Two parts of flower of sulphur, three parts of muriate of ammonia, and 16 parts of wrought iron filings, quite free from rust, are mixed thoroughly together and kept quite dry until required for use. When the cement is about to be used, one part of this compound is added to twelve parts of clean iron filings, turnings, or borings, in which those from wrought iron should preponderate, and the whole mixed with sufficient water (in which a few drops of sulphuric acid have been let fall) to bring it to a pasty consistency. In caulking sockets with this material it is necessary to exercise care, lest they are inadvertently fractured with the caulking tool, which should be selected of the proper size and cranked to enter the socket more readily. Further particulars connected with the use of this cement are given under the alternative title of Rust Joint.

Joggle Joint is one between cast iron pieces, having jogs or notches to prevent sliding. Large screw piles, for instance, with internal flanges bolted together, are strengthened against the strain of torsion at the joint by joggling the flanges.

Jump Joint occurs when the edges of plates forming a butt joint are squarely and smoothly planed so that the contact between them being perfect throughout, the cover plates and rivets are relieved of part of the compressive stress passing along the main plates, and are, therefore, respectively shorter and less numerous than with the ordinary butt.

Key Joint.—In screw-piling, a coupling joint consisting of a square or polygonal head fitting into a corresponding socket by which the whole can be revolved with capstan bars, &c. In some descriptions of work it is necessary to lash the outer extremities of the bars together to keep them in their holes in case of the capstan taking charge.

Lap or Lapped Joint.—In girder work, &c., this is the simplest form of riveted joint, and is made by overlapping the plates and uniting them by either a single, double, or triple riveted joint, under which terms further explanation will be found. Hoop-iron bond is lapped and riveted at angles. Sheets of galvanised iron are laid to form roof coverings with a lap of about $2\frac{1}{2}$ in. at the fluted and about 4 in. at the straight edges, and are secured with nails or screws to wooden bearers, or with bolts to iron ones.

Lead or Leaded Joint.—In bedding an iron column upon a post stone, bed stone, &c., a plate of sheet lead is now and again laid between the stone and iron base, which, by its yielding to the irregularities of surface, enables the column to remain vertical after it has been truly adjusted in that position. In the same way bed plates lie upon a piece of lead spread out upon the foundation slab. Other allusions to this practice will be found under Bed Joint. The operation of stopping joints with molten lead is described in the Plumbers' Section under Run Joint.

Lightning Conductor Joint.—This should be carefully hard soldered, and as few joints as possible used.

Longitudinal Joint is one running lengthwise. It sometimes occurs as a sort of butt joint in wide booms made up of two or three widths of plate instead of one only, the advantage of such a disposition being that the plates are longer and fewer covers required. In such situations the tensile or compressive strength of the boom is not affected by the longitudinal joint. In the manufacture of wrought iron piping, a perfect longitudinal joint, as well as a perfectly true pipe, is obtained by passing plates of iron at welding heat through properly prepared rollers. So sound is the joint thus made that the pipes required for steam purposes are often tested up to a pressure of 3,000 lbs. per square inch.

Mitre Joint.—An example of this kind of joint is noticeable at the junctions of the cross-bars of Harris's patent wrought iron windows.

Mortise and Tenon Joint is similar in principle to that made by the joiner. A caulking answers for the tenon, whilst the mortise is cut by the mason. In some localities the projection, or stub, or caulking left on the end of a casting for letting into masonry is called a tenon. Holes may likewise be punched in a piece of sheet-iron, and corresponding tenons cut on the edge of another and inserted therein, to be fixed with little rivets or keys. The framework of the panels of ornamental ironwork was usually connected together by means of the mortise and tenon joint during the sixteenth century.

Pin Joint occurs between tension bars or links. The diameter given to the pin, which is ascertained by formula, will depend upon whether it will be in double or quadruple shear, &c., and it must be such as to make it as strong in shearing resistance as the least section of the bar is in tensile resistance. The width of the metal round the pin-hole, and the distance of the centre of the latter from the end of the bar, have also to be calculated so as to obtain the necessary strength with the least possible weight at the junction. In some instances the pin is retained in place by a split key passing through a slot in the tail. Bars and booms of iron girders are connected by tight-fitting turned steel pins, sometimes as much as 7 in. in diameter, inserted through accurately-bored or drilled holes, or eyes. The pins are coated with grease and forced with considerable pressure into the holes, being kept therein by means of iron castings screwed on to both ends. The advantage claimed for this form of joint in girder work over a riveted one is, that the former admits of the girder being set up at the site exactly as when temporarily united at the works. It is alleged, however, and apparently with much reason, that pins are untrustworthy, since the long members of a bridge truss in compression jointed with them are circumstanced similarly to columns with rounded ends and susceptible of motion, or, in fact, are only one-third as strong as if their joints were made rigid by riveting. Interties between girders in fireproof floors are sometimes secured by surrounding the girder with a wrought iron band, saddle, or strap, and inserting the end of the intertie into a slot made in it at a point just above the lower flange, and securing it therein with a pin clinched with a split key.

Pipe Joint is a junction between two pipes without, as a rule, reducing their carrying capacity or fullway, and is usually made by means of a spigot and socket, or with flanges, or a coupling for direct extensions, but with the addition of a T-piece, or bend, or elbow for lateral ones. Flange joints have been already noticed. Cast iron pipes with spigots and sockets fitting loosely are usually run with lead, as described under Run Joint in Section X., and when made to fit tightly, as explained under Turned and Bored Joint. Red and white lead in equal quantities well mixed and caulked with white yarn is used for jointing wrought iron hot-water pipes. Other methods are described under the heads of India-rubber, Iron Cement, and Rust. When the pipes are large, collar joints may be employed, but the shorter the pipe the stronger the main, because the sockets act as bands. Besides the ordinary joint run with molten lead, another kind equally impervious is formed by inserting into the socket a ring of cast lead and driving the spigot end tightly inside it. Painter's patent hydrostatic pipe joint is an ingenious caulk made by squeezing a lead ring between the spigot and socket, which is of peculiar make, by means of the compression of a semi-fluid passed through a screw-hole in the socket, the force being supplied by a small screw-jack. Marini's pipe joint for steam, &c., consists of an iron collar between two grooved flanges, rings of vulcanised india-rubber being placed in the grooves and the flanges bolted through together. Service pipes when of iron are jointed with sockets screwed on the threaded end of each length with a touch of white lead, unless barrel unions likewise screwed are employed. Cast iron rain-water pipes are connected by inserting the spigot well packed with putty or red lead and tow into the socket, and their junction with the wall is effected with ears, or lugs, or clips and pipe nails. Before building into masonry, the soundness of the pipes should be tested by hydraulic pressure with a force pump.

Pivot or Pivoted Joint is a variety of hinge joint formed by a conical or cylindrical pin fixed to the posts of gates, or centres of rails, or centres or extremities of the hanging stiles of doors, &c., to enable them to swing freely. Dock gates furnish illustrations of its use, the heel posts of which sometimes work on steel pivots 12 in. in diameter let into post stones, the sills being supported by rollers working on cast iron roller paths or runners. A variety of swing bridge derives its name of pivot bridge from its moving on a vertical pivot situated midway between its two ends. Pivot lights are those supported at the sides by centre pins, otherwise pivots and centres.

Planed Joint is made by means of a machine tool called the planing machine, which to a great extent has superseded chipping and filing in producing, amongst other advantages, well-fitting, smooth, and true bearing surfaces, though planing does not eclipse chipping in producing evenness of surface. Through the instrumentality of planing, all the joints of ironwork, wrought or cast, can be finished off with parallel faces and a consequent precise and accurate fit and bearing, which has done much to extend the application of iron to structures and to inspire confidence in its resisting powers. The top and bottom surfaces of massive cast iron bed plates, the comparatively light ends of intermediate struts set between the segments of upper booms, and the butts and edges of plate iron from $\frac{1}{4}$ in. to 1 in. in thickness, are all well and smoothly surfaced by this ingenious process.

Punched Joint.—A riveted joint in which the rivet holes are punched and not drilled. As already observed, there appears to be a diversity of opinion amongst those well entitled to judge as to whether drilling or punching is most injurious to plates, but it is established that both operations at all events weaken the metal between the rivet holes as much as 16 per cent. Punching, however, which is the cheapest and clumsiest method, is more than suspected of causing a much greater loss of strength and to a greater distance beyond the rivet holes, especially since the drift or rimer has to be constantly used to get the rivets in. Of this increased loss, however, much is supposed to be recovered by the rivet head obtaining a greater grip through the operation of punching, whilst on the other hand the rivet in the drilled hole is more easily sheared whether it be left clean or ragged with the burr. Confining the comparison simply to the scope of these peculiarities, the balance stands slightly in favour of the punched joint. Plates should be punched from the side which forms one of the surfaces brought into close contact by the riveting, for the diameter of the rivet hole being less on the side on which the punch enters, and becoming a little larger as the disc of metal is forced out, the rivet obtains a better hold from being swelled out towards its ends. For this reason some recommend the necks of rivets to be bevelled under the heads. The necessary pressure for punching a circular hole of 1 in. diameter in $\frac{1}{2}$ in. plate-iron is according to some experiments about 26 tons, and according to others about 36 tons. Within certain limits indicated by this difference, the pressure varies directly as the thickness of plate and diameter of hole. The holes for attaching sheets of galvanised iron to roofs are punched by hand.

Radial Joint is a species of flange joint partaking of the nature of the ball and socket, explained in Section XIV., and allowing free motion within a conical space having a prescribed basal area. The contrivance has been taken advantage of in attaching columns to screw piles to allow for adjustment in case of the piles working out of position during the operation of screwing in.

Radiating Joint.—This occurs between cast iron solid and perforated voussoirs, segments, &c., and the sections of wrought ribs, &c.

Rebated Joint is made when a rebate is formed on a

casting in order that another piece may fit down into it to avoid an open joint, as is occasionally done, for example, in some kinds of cast iron sheet piling. Fig. 88 shows its application to lighter work where a continuous iron rail is rebated (or, perhaps, more correctly "halved in") and screwed together in lengths.

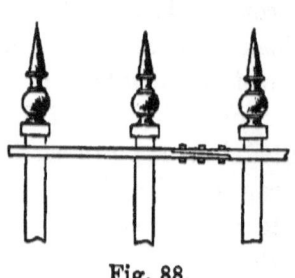
Fig. 88.

Red Lead Joint is one stopped and made tight with a mixture of red lead and linseed oil, or white lead may be added in the proportion of two parts white to one of red, or otherwise as preferred. Its use has been previously alluded to in this Section, and is further noticed under Screw Joint.

Riveted Joint is formed when plate, bar, angle, or T-irons, &c., are united by rivets usually made out of a tough and ductile iron, either scrap or a quality specially manufactured for them and known as rivet-iron, the tenacity of which ought not to be less than 22 tons per square inch of section. This joint is almost universal in wrought iron structures. Girders are formed of plates, bars, angles, &c., riveted together either directly or with joint and gusset plates, and with bent, cranked, and joggled stiffeners. Simple iron roof trusses are made by riveting the heads of the principals to a third piece or joint plate and bolting thereto the jaws of the suspending rod, and where the span is small, riveting the tie-rod also to the other members. In larger spans the members are built up with bars, angle, or T-irons, &c., riveted back to back and to one another, bolts being only used by rights for tensile strains, though they are not unfrequently put in posi-

Fig. 89.

tions where they are exposed to shearing. Angle iron cleats are much used for supporting purlins, and these should be riveted on before raising. A rivet consists of head and shank, or spindle, or tail, sufficiently long to pass through the parts to be joined and to be hammered down to a protuberance as large as the head. Good rivets when cold should stand doubling without showing a flaw, and the rivet holes ought to be spaced out with templets and be accurately bored so that they may exactly coincide when the pieces are put together. In hand riveting, a drift having been forced in and out again through opposite holes, the rivet, whose diameter when cold is about eleven-twelfths of the aperture's, is inserted red-hot from the inside

Fig. 90.

and a hammer immediately pressed against it, unless it can be laid upon an anvil or something handy and suitable. The plates are brought together by hammering round the rivet, which is clinched with a blow and finished off to a convex end, if so desired, with a riveting set having the proper concavity, whilst at the same time the shank is thickened out so as to leave no part of the hole empty. Twenty minutes is the longest time 1 in. or ⅞ in. rivets should be in the fire, and all showing signs of burning must be rejected. Loose rivets are detected by tapping, but the detection is not so certain when the holes are punched. The operation of riveting is now frequently carried out by a steam or hydraulic machine where it can be easily applied, but not in awkward situations. Sometimes cold rivets are used instead of hot where it would be impracticable to manipulate the latter. Rivets have different names according to the shape imparted to their terminations. Fig. 89 represents a button or cup-shaped snap rivet, or one to which a round and symmetrical shape has been communicated before cooling by means of a tool called a snap, having a hollow face of the proper concavity. Fig. 90 shows a hammered or staff rivet; and Fig. 91 a countersunk one, useful in obtaining close bearing surfaces. The heads also vary both in name and shape, but mostly in name, being styled either flat, pan, cheese, button, cup, or rose shaped. As a rule, their diameter equals about two thicknesses of plate when less than ½ in. thick, but when that or

Fig. 91.

more it equals only 1½ its thickness. The height of head should not be more than half the diameter of shank, and the diameter of head should about equal two diameters of shank, and the length of the shank exclusive of head before clinching equals the sum of the thickness of the plates to be joined, plus a length equal to about 1½ the diameter of the rivet for the tail or point. The contraction of the rivets in cooling produces a grip which adds considerable frictional force to the strength of the joint, which, however, in double and single riveting, owing to the abstraction of metal for the holes, is usually from 20 to more than 30 per cent. less than the strength of the plate, which itself is sometimes reduced nearly 20 per cent. through clumsy punching and rough usage during the operation of riveting. Since it is impossible to paint rivets and rivet-holes and joints where the metal is in contact, oxidation is assisted in exposed structures at these parts by the incessant vibration occasioned by wind and moving loads, and any protective film, such, for instance, as that of black oxide produced by Barff's process, would be lost in forcing rivets into their holes. In practice, a riveted joint fails either by the crushing of the plates between the rivets and edge, which rarely ever occurs, or by the tearing of the plate along the line of rivets between the holes, or by the shearing of the rivets; and these three modes of fracture respectively depend on the thicknesses of plate and rivet, the pitch of the rivets and thickness of plate, and the rivet's thickness. The quality of the iron used, even if occasionally too hard, is scarcely ever the cause of disaster, but the same cannot be averred as regards the quality of workmanship. Various kinds of riveted joint are noticed under their specific names.

Run Joint.—Small wrought iron columns cut from gas tubing are sometimes united to cast iron sole plates by running with molten iron in preference to lead. Iron cramps used for uniting blocks of masonry put in hot, and run and completely surrounded with asphalte, are apparently secure against oxidation. Cast iron is much safer than wrought, unless the latter is galvanised where there is danger of the enveloping coat being abraded. It must be admitted, however, that on this point opposite views are entertained, owing to all iron, and especially cast iron, corroding by pitting, and thus becoming perhaps honeycombed or eaten into holes in the interior, whilst its visible surface or skin is barely affected. Sulphur is not much in vogue in this country for running joints, its great advantage being that it expands in cooling. When it is employed, only just sufficient heat must be applied to melt it, otherwise it will run with difficulty, and it must not be allowed to burn. The surplus should be removed before it gets cold enough to become brittle, and sometimes dry sand is mixed with the sulphur with the object of hardening it. Ironwork is not dangerous when let into brickwork, but it is so in masonry without a protective coating, which in cast iron perhaps more or less effectively exists in its own skin.

Rust Joint.—This is so called on account of its enveloping the parts joined with an air-tight mass of rust, which becomes sounder by age. It is made by caulking the flanges of ironwork and the sockets of iron pipes previously partially packed with two or three rings of spun yarn, with a composition called iron cement, or rust cement, consisting of iron filings, borings, or turnings, and sal ammoniac mixed together in different proportions with or without sulphur. A good variety is described under Iron Cement Joint. The proportions of the components, however, vary in different receipts, one recommending sulphur, sal ammoniac (muriate of ammonia), and iron borings, in the respective proportions of 2, 1, and 80. These are to be separately powdered, and then well mixed together with water sufficient to moisten them into a paste, which will oxidise and swell and set on exposure to the air. It requires careful preparing and caulking also, otherwise the socket will be in jeopardy of bursting. The joint between hoop iron and mortar or cement is greatly strengthened by slightly rusting the iron.

S-Joint.—This is similar to the corresponding joint of the Zinc Worker.

Saddle Joint.—Besides the joint of the same name described in Section XI., and which is applicable to this trade also, there is another important connection occurring between the chains of suspension bridges and the saddles on which they ride that from analogy rightly goes by this term. The saddles on the piers carrying the chains of the Clifton Suspension Bridge are of wrought iron, placed on roller frames of cast iron, the rollers being of cast steel. In piping a similarly-named joint results from saddling a main with an iron band composed of two or more sections which are bolted together and caulked. The main is then retapped and the service thus becomes saddle-jointed.

Sand Joint is formed by turning down the edges of a lid or cover and making them dip into a continuous groove filled with sand. It is nearly air-tight, and is used with some kinds of stoves to render the lid of the outer casing smoke-tight.

Screw Joint is made by means of a screw, which is a cylinder having its surface cut so as to leave a projecting spiral ridge, thread, or worm, whose inclination to the axis is constant. The screw is turned into a hole drilled and tapped to an exact counterpart and called a female screw, the tapping not being required, however, in the case of wood screws or screw nails, owing to the facility with which metal threads tap their own channels in most kinds of wood. Before the introduction of the socket joint the parts of iron mains used to be screwed together—a practice which is said to have led to some breakage through hindering expansion and contraction, which it must be observed cannot be much with properly interred pipes. Mains, indeed, are supposed to go with the ground, and are therefore kept low enough to avoid the chilly influences of frosty air, but in these instances, doubtless through imperfect threading and bedding, the screw ends were unequally strained, which sudden changes of temperature sufficiently augmented to produce fracture. Lengths of wrought iron barrel are united with screw sockets and red lead cement with or without a thread of yarn. In order that a screw may be as strong against the shearing off of its thread as against its rupture by direct tearing asunder, the depth of the overlap of the male and female threads should be theoretically about one-half the least effective diameter, and in practice this is doubled, as mentioned under Bolted Joint in Section VI., the depth of the overlap being made equal to the diameter.

Screw Shackle Joint.—One form has been already described under the term adjustable, and its nature will be found immediately below under the head of Shackle Joint. The form of the shackle is of little consequence so long as it admits of regulating to a nicety the length of the tie whilst it holds the halves of the rod, which it tightens up, with sufficient rigidity. Fig. 92 represents one of many forms. Tension rods to princi-

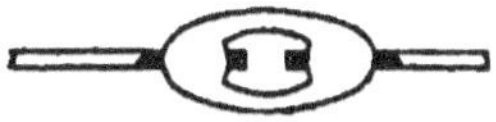

Fig. 92.

pals and tie-bolts to strutting to floors are provided with screw shackles for adjustment; and screwed ends at all points of connection of sections of tension rods, when the latter are of considerable length, are advisable in lieu of welds.

Shackle Joint is made with a shackle which, if not worked by a screw, consists of some such contrivance as a curved or stirrup-shaped bar, through whose extremities pass a single wedge, or a pair, in order to limit and control the tension of the tie-rod, &c., and consequently the free action of the piece it keeps in place. It is often omitted, however, and the consequence is that suspension rods, tension rods, and diagonal ties, &c., are either too slack or too much strained. The shackles used in building are composed as for other purposes, of a more or less simple arrangement of links, pins, and rods, but in order to admit of adjustment they must be provided with screws or cotters.

Shoe Joint.—This is formed by means of a cast iron shoe or chair, which is a medium of uniting beams, or rods, or rafters to walls, columns, or overhead girders, and struts to suspending rods, &c. It constitutes as characteristic a joint as that produced by a hinge, and singularly enough their analogy or affinity is even closer than that arising from the similarity resulting from the interposition of a metallic body between the parts connected, for shoes themselves are sometimes hinged when fixed over columns supporting railway bridges, &c., in order that the girders may transmit their load down the axes of the columns. Other remarks concerning shoes have been included in the description given of Cottered Joint.

Single Riveted Joint is made when two plates are united by a single row or file of rivets. The joint may be a lap between two main plates, or a cover in which each main plate is attached to one or two cover plates by one file only. In a single riveted lap joint in which the rivets have a diameter equal to two thicknesses of the plate, and where the lap and pitch are both three diameters, which is the best arrangement for uniformity of strength throughout the joint, the strength of the joint is only 52 per cent. that of the plate. A single riveted butt joint with a cover plate on only one side has likewise the same proportionate strength, but when there are two cover plates the rivets offer a double shearing area, or, in other words, are in double shear, and their diameter can accordingly be reduced to $1\frac{1}{8}$ the thickness of the plate, the pitch to $3\frac{1}{8}$ diameters, and the lap to 3, in order to obtain similar equivalent strength.

Socket Joint.—The variety formed by the union of a spigot and socket is described under that name and Pipe Joint as well. The tops of iron piles carrying columns are provided with sockets of octagonal form exteriorly for screwing in, but with circular apertures for receiving the feet of the overhead columns, which are secured therein with iron wedges and iron cement or other approved material.

Soldered Joint.—Soft and hard solders are used for iron as well as for metals generally, with or without rivets, fractured cast iron even being made in some instances as strong as ever if properly tinned. For further particulars respecting the application of soft solder to iron see Copper-bit Joint in Section X. Hard soldering is sometimes effected as follows:—The surfaces to be united are filed clean, and, if heavy, bound together with steel, a thin strip of sheet copper or brass being laid over the joint and tied on with wire. A band of clay, free from sand, about an inch thick and 10 in. wide, is placed over the joint, and the whole laid near a fire, so that the clay may slowly dry, after which it is held before the blast and brought to a white heat. It must be allowed to cool slowly if the junction is between steel pieces, but on the contrary if between iron ones cooling off in water is necessary. The vitrified clay is finally broken away and the surface cleaned off. See also Brazed Joint, Section XII.

Spigot and Faucet Joint.—This is the same as the spigot and socket joint.

Spigot and Socket Joint is formed by inserting the spigot or straight end of one tubular piece into the widened-out end called the socket or faucet of another piece usually of the same diameter. In piping, each piece invariably has a spigot and socket at its opposite ends, the spigot, however, being in many cases nothing more than a plain end. Cast iron piles are sometimes thus united, the lower length being first fixed and the upper part lowered into the socket and strongly bolted thereto. It constitutes a common means of connecting pipes, the part of the socket not occupied by the spigot being filled with some kind of stuffing, as described under Cement Joint, Run Joint, &c. Even when these ends are turned and bored, the joint is all the sounder for the insertion of a little red lead paint. The spigot and socket joint, properly made, preserves unimpaired the fullway or complete carrying capacity of the pipe. In some instances rings or strips of cold lead about $\frac{3}{4}$ in. wide have been driven into the sockets of cast iron pipes previous to caulking with white yarn and molten lead, but the practice is not usually adopted. Occasionally for water mains the sockets have been securely wedged up with wood taper plugs of proper form to fit the annular cavity.

Strut Joint.—Heads of wrought iron struts in iron roofs are secured by riveting a flat strip or piece of iron on both sides of the web of the strut, the two pieces embracing likewise the web of the rafter to which they are also riveted or bolted. The foot of the strut if of T-iron is attached to the tie-rod by bending the table and bolting through the flattened-out tie-rod with double nuts and bevelled washer. Cast iron struts are sometimes terminated with forks or jaws to grip the rafter, and sometimes with a circular box which holds the knobbed ends of the tension and tie-rods, an ornamental boss or cover being added after the ends have been inserted. Struts formed out of wrought iron gas tubing have cast sockets attached for bolting to the other parts. I girders particularly require strutting to give them the proper lateral rigidity, and so do all wrought iron members when fixed to resist compression, the intersecting joints of cross-bracing often furnishing the appropriate means. In some cases, as for instance in bridge work, the struts of the bracing should be well wedged up and washers inserted upon the withdrawal of the wedges previous to completing the joint.

Sulphur Joint is one between metal cramps, corkings, &c., and stone, the cementing medium being sulphur, which is fused and used as described under Run Joint. One great advantage in thus employing sulphur arises from its expanding in cooling, whereas lead contracts and requires to be caulked. The use of sulphur in iron cement is alleged to expose the sockets to fracture through an expansive power which it imparts to the cement. Iron sulphide is very hard, and whilst the efficiency of the cement is doubtless owing to its formation as well as to the expansion of the iron in combining with the sal-ammoniac, practice demonstrates that unless carefully prepared the cement may in time burst the sockets, though, as elsewhere suggested, clumsy caulking may give the initial crack to the disaster, which would perhaps not occur with a faultless socket of uniform thickness.

T-Joint is formed by uniting pipes or plates by means of a T-piece shaped like the letter T. A gas main must be bladdered, and a bye pipe inserted before a portion of the main can be cut out to introduce a T for a large service, since the supply of gas cannot be even momentarily cut off with the same impunity as water for a much longer interval. By cutting out round the main a deep groove with a round-nosed cutting punch at the two ends of the length to be removed, it will easily part with a few blows of the hammer, and after the piece is taken out a collar is slipped over the main to cover the butt between the plain ends of the main and T, whilst the joint at the other end of the T-piece is made in the usual way by inserting the spigot into the socket.

Thimble Joint is a form of expansion joint between lengths of pipe with butt ends which, being turned true, are encompassed by a narrow strip of tin, and this again is enclosed by a metal collar or thimble. Between this thimble and the tin strip, packing is stuffed to insure a tight joint, whilst permitting the ends to separate and approach during changes of temperature.

Through and Through Joint.—A variety of butt otherwise called a full butt.

Tie Joint is made when the end of any part or piece in tension is secured to its support. Tie-bolts which are usually made out of 1 in. round iron securing concrete floors, &c., either pass through the wall and are held by anchor plates or cast iron discs, to which they are screwed, or else their ends are flattened and turned down and rest against a wrought bar 3 in. or 4 in. wide by $\frac{1}{2}$ in. thick, built purposely into the wall to receive them. The bar, however, is often dispensed with. In timber partitions the end of the rod is usually retained in place by a nut and large washer. Tie-bars are riveted or bolted, as preferred, to the joint plates, &c., of girders. Horizontal bracing between girders, tension rods, diagonal ties, &c., are bolted or riveted at one extremity, and adjusted with cottered or screw shackle joints at the other. The chairs at the feet of rafters and the heads at their apices form convenient holds for the ties of a truss. The tie bolts of cofferdams require to be very stout, and are secured with nuts and large washers. Girders may be tied to walls with horizontal straps bolted to the former, and built into the latter, or else vertical bolts may be built up with the masonry, or lewis bolts, or dowels even, may be leaded to the stone templates. Provision for expansion is not necessary when the span is less than 60 feet.

Tight Joint, in piping or allied work, is one that is impervious to air or fluid under any pressure likely to be applied to it, and the method of obtaining it has been described under previous headings. In the construction of cofferdams, which have been aptly defined as water-tight walls, the joints of iron piling are usually made impenetrable to any abnormal ingress of water by careful fitting and fixing and a backing of concrete. The subject is further noticed under Water-tight Joint.

Triple Riveted Joint is an arrangement of rivets in three rows instead of in two as in double riveting, by which means the rivets can be spaced at greater distances apart to obtain a joint equally strong throughout. Considerable increase of strength amounting to about 20 per cent. results from its adoption, and more still from that of quadruple riveting, but the difficulty of caulking without springing the plates is augmented as the pitch increases, and for structures where this process is imperative it becomes a good argument against adopting this sort of joint. In a grouped riveted joint in girder work this species of multiple riveting, or a modification of it, is frequently met with.

True Joint.—This occurs when the surfaces joined are close, smooth, and parallel throughout.

Turned and Bored Joint is used in jointing cast iron pipes, and consists of a spigot and socket accurately turned in a lathe, the spigot with a slight taper and the socket a little cone-shaped to exactly fit it. The ends are then smeared with whiting and tallow, or painted with a mixture of red and white lead or other suitable compound, and when put together the joint is perfected by forcing the spigot well home with a mallet or otherwise. This method is only adapted for pipes laid in straight lines or in curves of considerable radius, otherwise the ordinary lead joint or some other admitting of a little cranking must be employed. It has been used, however, with pipes up to 48 in. in diameter, and possesses the advantage of not requiring so wide a trench as when the main layers have to get all round the sockets. A form of turned and bored joint is also found in pinned attachments occurring between the booms and bracing bars of girders where the pins are turned and the holes accurately bored to their exact size.

Union Joint is a mode of uniting the ends of pipes, rods, &c., by means of a coupling, which for pipes will be found in Section XIV. Unions for connecting rods have been explained under Screw Shackle, which embodies the principle of all those most in vogue for tension rods.

Universal Joint.—This is the name given to a con-

Fig. 93.

trivance for connecting two rods or pipes so that each may have freedom of motion in any direction with certain restrictions as to extent. Fig. 93 is an example admitting great range of motion, the forked ends of two rods being pivoted to the extremities of two crosses which are likewise pivoted to a short intermediate connecting piece.

Vertical Joint occurs between the parts of cast iron rings or cylinders when built up in segments, and likewise at the junctions of iron sheet piling, &c., &c.

Water-tight Joint.—In wrought iron cofferdams the joints between the cylindrical pieces are rendered water-tight by driving feather piles of timber into opposite recesses provided in the metal surfaces to receive them. The joints between the segments also of cast iron columns constructed in rings, and used in submerged foundations, and those of other cognate structures, are rendered water-tight with red lead, the surfaces being properly faced. Beeswax, asphalte, or other non-porous and adhesive substances are useful in obtaining impermeability between rougher surfaces. Caulking with iron cement is the common method of making the joints of castings water-tight, whilst a liberal application of tar answers the same purpose when wrought iron work is to be insured against percolating jointings. Messrs. Burt and Potts have introduced a patent wrought iron water-tight casement and frame, suitable for very exposed situations.

Wedged Joint.—The use of the wedge in securing joints is very commonly met with. Cottered joints owe their adjustability to wedges. Columns are wedged up plumb before running their bed joints. Overhead columns are thus tightened up in the sockets or caps of piles, and in many cases iron piles are secured to holes bored in rock with wrought iron wedges, with or without concrete. Small solid wrought iron bars, forming the legs of piers, &c., have been made fast to foundation holes cut in the rock to the proper size and tapered at bottom by bringing the bars first to a white heat, then splitting their feet and inserting wedges therein. Thus prepared, the bars were immediately dropped into place and driven home before the ends became too hard to assume the requisite enlargement.

Welded Joint.—In theory the welded joint is unsurpassed in strength and thoroughness. It results from cohesion between the two surfaces coming into play, the white heat enabling compression or a single vigorous blow to bring the opposite particles within a sufficiently insensible distance to cause the law of cohesion to at once operate, whereby the joint is, as it were, lost in the entirety of the metal. Every smith cannot make a good weld, hence the utmost firmness should be shown in rejecting iron fastenings suspected of any treacherous flaw, or of being badly forged. Before being raised to the welding heat, which is just above whiteness, the two pieces to be united are roughly shaped and scarfed at a bright red heat so as to overlap and approximately fit, and when both are white hot sand must be sprinkled over the surfaces about to be brought into contact, for by fusing and spreading it prevents oxidation and retards the combustion of the metal. If the joint is too long to be completed at one blow the work must be returned to be reheated, and after the union is thus quite effected further blows will tend to consolidate the joint and bring the work to the desired shape. Although the process appears tolerably simple, perhaps it is no exaggeration to say that a sound weld between two plates is unobtainable, and between bars, rods, &c., owing to the imperfection of workmanship, always uncertain, therefore the more important the work the more it is avoided, but there are occasions, nevertheless, when the welded joint if skilfully executed proves both useful and safe.

Zigzag Riveted Joint.—This is a term applied in riveting when the rivets in each parallel row, or file, or pitch line, are spaced so as not to stand immediately behind one another in the line of stress, but so that each rivet is opposite the centre of a space between the rivets of any adjacent line, as in Fig. 94. Consequently a line joining their centres would run from right to left in zigzag fashion. To obtain the strongest lap joint with two rows of rivets thus arranged the distance between the pitch lines ought to be about two-thirds the pitch, which may be made equal to $4\frac{1}{2}$ or 5 diameters.

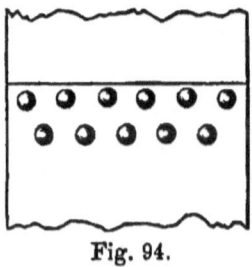

Fig. 94.

SECTION VIII.

JOINERS' JOINTS.

Angle Joint is one occurring at an angle. A mitre is pre-eminently an angle joint, but apart from those thus distinguished there are many others with specific names, such, for instance, as bead and double quirk, Fig. 95, which is common in salient or external angles.

Fig. 95.

Angular Joint.—This contains an angle, or is made with an angular piece. The edges of "grounds" bevelled to receive or key the plastering illustrate this kind of joint, which is formed between surfaces whereof one at least is inclined to the face.

Angular Grooved and Tongued Joint.—This is made by sinking an angular groove and working a corresponding tongue to match. With ordinary stuff it is advisable to keep the groove well within the arris, as shown in Fig. 96, but in mediæval times the wood used was sufficiently well selected and tough, and the workmanship laborious enough to admit of the groove occupying nearly the whole width of the edge, as in Fig. 97.

Fig. 96.

Fig. 97.

Beaded Joint is one disguised by a quirked bead, which is worked on one side of it, as in Fig. 98, in such a way that when well done the joint appears as a second quirk. It is common in matched boarding, hinging, &c.

Broken Joint.—Contrary to what is the case in the other trades, this kind

Fig. 98.

of joint is not regarded with particular favour, for it occurs in patched and the commoner descriptions of work, notably in floors laid folding in which there are breaks in the longitudinal joints arising from the floor boards being gauged to different widths.

Butt Joint.—This is formed when two boards with edges shot or ends squared, and either glued or not, abut against or touch each other edge to edge, edge to end, or end to end, but custom gives it a still wider meaning, and the term is applied to the joint made when the edge of one board either plain, beaded, or tongued, is fitted down to the side of another either recessed or not, as in Figs. 99, 100, and 101. In handrailing it occurs when the plane of the joint is perpendicular to the direction of the rail.

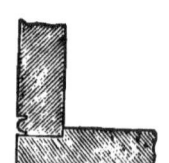

Fig. 99.

Fig. 100.

Chamfered Joint is made by taking off the arrises of boards, panels, and stiles which butt against each other, thereby leaving a **V**-shaped recess of some little effect.

Fig. 101.

Checked Joint.—This is identical with the similarly-named carpenters' joint.

Circular Joint is found in circular work, as, for instance, at the shutting joints of circular or segmental headed or pointed sashes and casements, and in circular panelling, &c.

Cistern Joint is a sort of grooved and tongued joint, the groove being worked on the side of the board. When dovetailed, as shown by the dotted lines in Fig. 102, the tongued board must be slipped into place endways.

Clamped Joint.—This is effected by joining the ends of boards already glued up with a piece called a clamp, which is secured by means of a groove sunk in it and a tongue cut across the fibres of the boards, these being often tenoned as well to fit into mortises in the clamp. In other instances the clamp is tongued and the boards cross grooved. The mitre clamped joint occurs when the ends of the clamp are mitred, as shown in Fig. 103. The object of the clamp in each case is to prevent the boards from casting and to give a more finished appearance. A joint between floor boards is also clamped when forced close with a flooring clamp or cramp, these two terms being now interchangeable and constantly used the one for the other.

Fig. 102.

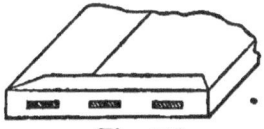

Fig. 103.

Close Joint.—This signifies one without any appreciable gape or crevice. Floors are laid with close joints, and in good work they should not open $\frac{1}{16}$ in. within a year of being laid. The same rule applies to all other joiners' work not glued up, and where the deals used have been quite dry there will be much less shrinkage than that named within the same period.

Cramp or Cramped Joint.—This is one tightened up by means of a screw cramp or clamp, and it is to be observed that the terms cramp and clamp are and may be indiscriminately used for the tool employed to compress joints as well as for a piece fitted to prevent casting as above described.

Cross Grooved and Tongued Joint is made when a groove is cut out across the grain, as in heading and cistern joints, &c.

Double Lapped Joint.—This is adapted for hinging together two stiles, and is illustrated in Fig. 104.

Double Quirk Bead Joint.—This is shown in Fig. 95,

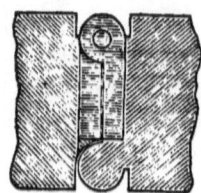

Fig. 104.

but it need not be grooved and tongued, nor must it of necessity be formed at an angle. When it is, and the angle happens to be a right angle, the bead is three-quarters of a cylinder, but when the angle is obtuse the bead is less. By beading matched boarding, rebated boarding, panelling, &c., the visible joint becomes a bead and double quirk, which is likewise well shown in Fig. 104.

Double Tongued Joint is shown in Fig. 105. The

Fig. 105.

same principle can be extended easily to more complex sections, but as a rule in the matter of joints utility and complication are in inverse proportion.

Dovetail Joint.—This is not much used in joinery excepting for dovetailed backings to door linings, &c., shelving, drawers, &c. It is necessary that the fibres of pieces thus united should run parallel, so that pin and socket may contract or swell equally. In the common dovetail the dovetail appears on both sides of the angle. Fig. 116 shows the lapped dovetail.

Dowelled Joint is formed by placing the meeting edges of two boards true and square, with iron or hard

Fig. 106.

wood pins or pegs called dowels inserted between them, as shown in Fig. 106, at equal-spaced intervals and in the middle of their thickness, to keep the face of the work in the same plane, or from buckling. In flooring, however, the dowels are sometimes put in low to avoid disturbance through wear. Dowels are likewise employed in other positions, as, for instance, to strengthen the intersections of window bars.

Feather or Feather Tongued Joint.—This is made by grooving the butting edges of boards shot true and square, and joining them by inserting and glueing a slip feather tight into the grooves, as represented in Fig. 109. A slip feather may either be a slip of harder and tougher wood, or else one that is cut out somewhat across the grain. A strip of wrought iron slipped into the opposite grooves is now often called by the same term.

Feather Wedged Joint is the same as fox wedged.

Fillistered Joint.—This is another name for a rebated joint, and so called because the rectangular recess termed a rebate or rabbet, Fig. 107, is readily made of any depth or width with a plane called a side fillister, having a vertically sliding fence, with a screw stop on one side which determines the depth of the rebate, for it will not allow the plane to descend further than the distance to which it is set. On the bottom of the sole also there is a shift-

Fig. 107.

ing or adjustable fence fastened by two screws working in slots, by which the width of the rebate is limited to the extent desired. In front of the grooving or cutting iron there is a cutting point to divide the fibres longitudinally and enable the cutting iron to preserve a clean and square sinking.

Folding Joint is similar to a rule or hinge joint.

Fox Wedged Joint.—This is the same as that of a like name described in Section VI.

Framed Joint is made with a mortise and tenon.

Franked Joint is that produced by the operation called franking, which is sometimes performed at the intersections of the cross or horizontal with the upright or vertical bars of a sash, in order to leave the through bars as strong as possible. In common sash windows the vertical bars extend from rail to rail, but in casements, so as to better withstand the jar, the cross bars are continued through in one piece. The others have consequently to be cut, fitted, and tenoned into the continuous or through ones, and when in order to effect this their moulded sides have to be cut into, the part notched out therefrom is called the franking, whilst the term franking also signifies the performance of the work. The tenons meet in the centre of the mortise, and are sometimes dowelled. There are other less complicated and perhaps equally sound methods of uniting the bars. One of these is to cut a thin tenon in the shorter bar and scribe the moulding of the through bar, as shown in Fig. 133. Another plan, but one more likely to produce a faulty bar owing to notching being necessary as well as mortising, is to mitre all the mouldings at their intersection.

Glued or Glued Up Joint is one in which glue is used between the edges of boards, &c., to unite their surfaces or to impart additional strength to the junction. It is effected by shooting the edges perfectly true, cleaning and warming them. They are then thinly coated with hot glue, and rubbed together until almost all of it is forced out and what is left is nearly set. The motion of one piece on the other need be but slight—an inch or two each way only—and as soon as they feel inclined to stick they must not be moved again. As a general rule, all glued joints should be feather tongued, but glueing is used as little as possible in good joinery excepting in the joints down panels, the untrammelled groove and tongue affording better opportunities for shrinkage and the display of tasteful mouldings. Almost all small mortises and tenons, after being fitted and trimmed and left as long as possible, are put together with glue and wedged, and if the glue used be of the best quality it is but little affected by atmospheric changes. The carpenter, of course, never uses glue, and as to joinery, whatever assistance it may lend to good work, no doubt glue is a fast friend to bad work. In external work white lead replaces glue.

Glued and Blocked Joint.—This is the ordinary glued joint strengthened by a block or blocking of prismatic form, and is common in stairs, built-up mouldings, cornices, columns, &c. Fig. 108 shows the joint between a tread and its rebated riser stiffened with blocks of triangular section, which discharge the useful function of checking disagreeable creaking.

Fig. 108.

Grooved and Feathered Joint is formed when uniting two boards with a slip feather of hard or obliquely-cut wood or iron tongue inserted, as in Fig. 109, in corresponding grooves, which are run by means of a plough, as explained under Ploughed and Feathered Joint. The same term is, however, frequently used for the common grooved and tongued, or ploughed and tongued joint, where the tongue is not detached but worked on one of the boards, as noticed under Grooved and Tongued Joint.

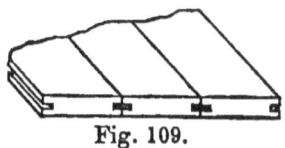

Fig. 109.

Grooved and Rebated Joint is made, as shown in Fig. 110, by combining a rebate with a groove and tongue. It offers a superior hold for edge-nailing when adopted for flooring.

Fig. 110.

Grooved and Tongued Joint.—This is much in vogue in joinery for stopping rays of light and the passage of air and dust, and is invaluable for allowing shrinkage to occur without sacrificing effect, provided a moulding of some kind is struck or planted on either side of the seam. The joint is shown in Fig. 111, and is made by sinking with the plough on the edge or side of a board a rectangular cavity, or recess, or groove, and working on the edge of another with a rebate plane or the side fillister—by taking a rebate out of each side of the edge—a projecting fillet or tongue that will exactly fit the groove. In the case of floor and other boarding the groove and tongue are so situated as to retain the surfaces in a true plane. This joint is also called ploughed and tongued, grooved and feathered, and ploughed and feathered, but it is more correct to apply the last two terms only where iron or slip feathers are used.

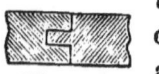

Fig. 111.

Grooved Tongued and Beaded Joint occurs as shown in Figs. 95 and 98 when a quirked bead is stuck on the tongued edge of a board, so that when shrinkage causes the tongue to draw a little out of the groove the opening and its shadow may look like a second quirk or part of the moulding.

Grooved Tongued and Mitred Joint is best explained by reference to Fig. 112. It possesses the qualities of strength and effectiveness.

Half Mortise and Tenon Joint is a kind of mortise and tenon whereof the mortise is cut through or has only one side. It is common in the heads of door and window frames.

Hammer Headed Key Joint.—This is sometimes found in the heads of circular window frames, the radiating joints being tightened up and held together by means of a key of hard wood with both extremities thickened out, and somewhat resembling in shape the head of a framing hammer.

Handrail Joint.—Usually a butt joint secured with a joint screw, otherwise handrail screw, which has a right and left hand thread each working in a nut with notches, both the nuts being buried in holes sunk near the ends of the two lengths of rail. By turning the nuts the joint is tightened up, after which the holes are stopped with pieces of wood to match, and the mouldings of the handrail finished off with handrail planes. Sometimes, however, the connection is made with a splice joint, which is a species of scarf, the planes of junction being horizontal and vertical when the rail is in position.

Heading Joint is formed between lengths of handrails, and of boards, &c., connected end to end, and is either square, cross grooved or tongued, splayed or bevelled, rebated and tongued, or forked. In dados and similar glued-up work it is glued.

Hinge Joint.—A general term for many joints that turn or, in other words, that open and close by means of ligaments of any degree of plainness and elaborateness called hinges (*alias* gemmels in mediæval times) which are fixed on or near the surfaces of contact either to bring them together or to part them as desired. Hinging and hanging are with the joiner almost correlative terms. To make a door swing easily and not expose the inside of a room when open whilst it will shut close without sticking, is a good test of a good workman. The whole subject of hinging is so enticing and interesting that it will be expedient to pass on at once to the next joint.

Hook Joint.—This is formed when the rebates of the meeting stiles of a casement are worked out with a curved groove so as to interlock, as shown in Fig. 113, for the purpose of keeping out rain. Two fillets, however, fixed inside and outside to opposite stiles is a more secure arrangement, whilst a single one on the outside, ornamented with a cocked bead, is often adopted and deemed sufficient.

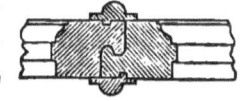

Fig. 113.

Housed Joint.—This is the same in principle as that made by the carpenter, but there are a few points of variation that require noticing. Instead of the whole or nearly the whole end of a piece being housed in, as is usually the case when that artificer makes a joint of this description, or indeed as happens when the joiner himself lets skirtings into architraves, blocks, &c., there are occasions in joinery of frequent occurrence that necessitate only about half and not the full thickness of a moulded board, &c., being housed in, as when skirtings are housed at a re-entering angle. On the other hand, sometimes more is hollowed out of one board than is merely required for the thickness of the other, as, for example, when steps are housed into strings, for here the grooves or housings are cut large enough to take the glued wedges also. Again, the *edge* of a board for its whole length and full thickness may be housed into the face of another, as instanced by the joint not unfrequently made between the inside lining of a sash frame and the soffit of the window lining.

India-rubber Joint.—A strip of this material is occasionally used as a weather joint on a casement sill.

Key Joint.—This arises in joinery when dados, &c., are keyed by means of stiff tapering pieces dovetailed, but not glued, into cross grooves, to allow shrinkage and extension; and again when, in other cases, short slip feathers or small pieces of hard wood of different shapes, likewise termed keys, are set with glue into saw kerfs, &c., to help to bind connected parts together. Hammer-headed key joint and mitre joint keyed are illustrations of this variety.

Keyed Joint is an alternative term for key joint.

Keyed Mitre Joint is another name for mitre joint keyed.

Lap or Lapped Joint.—Figs. 104 and 114 represent examples of lapped hinge joints. Feather edge boarding, Fig. 115, is an instance of the joint when one edge of a board is thinned down and made to taper.

Fig. 114.

Fig. 115.

Lap or Lapped Dovetail Joint.—A form of joint in which the dovetails appear only on one side, as in Fig. 116, the holes in the other side being stopped short of the face.

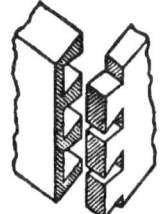

Fig. 116.

Lapped and Mitred Dovetail Joint.— To all appearances this joint is merely mitred, the pins and sockets of the dovetails being concealed, as they are only cut through about two-thirds of the thickness of the boards, as shown by the dotted lines in Fig. 117. The projecting lap on each board is mitred all the way down, and the edges at the top and bottom, or at all events at the top, mitred right through.

Fig. 117.

Lapped and Tongued Mitre Joint is made occasionally between boards, and is formed by rebating them both, as shown in Fig. 118, mitring them slightly at the angle, and keying them with a cross tongue.

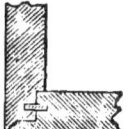

Fig. 118.

Lapped Mitre Joint is used for joining boards of the same or a different thickness, and is made by rebating one of them, mitring only a small portion at the angle, and securing with nails. Fig. 119 gives a section of it.

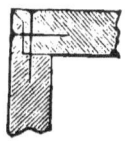

Fig. 119.

Lateral Joint in joinery occurs between the edges of boards, and has various specific names, of which two of the commonest are, perhaps, the square edged and the grooved and tongued. It is the same as a longitudinal joint.

Longitudinal Joint.—This runs parallel to the fibres of the stuff, and is well exemplified in the seams between different descriptions of boarding and lining.

Mitre Joint.—The common mitre is made by bevelling

off the edges of each board to an angle of 45°, and butting the edges together with the arrises in contact as in Fig. 120, but for any other variety of mitre the boards need not necessarily be of equal thickness, and the cut edges may be of any obliquity, the only condition being that each piece must be cut to the same angle. The common mitre is constantly met with in joiners' work, particularly at the intersections of window bars, the corners of square panelling where mouldings have been planted in, and also at the salient angles of skirtings, borders to hearths, &c. In order to bring the ends of the pieces to a proper angle to make a good mitre, the joiner uses a tool called a mitre block, or box, the latter containing vertical saw kerfs or grooves intersecting at an angle of 45°, and the work being closely pressed against the box, the saw is almost necessarily made to effect the requisite clean and oblique cut. Mitre shoots are also used to guide planes in smoothing off the bevelled surfaces.

Fig. 120.

Mitre and Butt Joint occurs as in Fig. 121, when two boards of different thicknesses are united together by bevelling the thinnest throughout the entire edge, and the thicker of the two to the same extent and angle, whereby a combination of the butt and mitre is produced. It presents facilities for nailing, and no opening in shrinking externally appears.

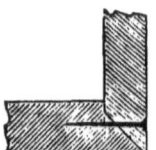
Fig. 121.

Mitred and Cross Tongued Joint is synonymous with mitre joint keyed.

Mitre Clamped Joint is made when a clamped joint is mitred as in Fig. 103.

Mitred Dovetail Joint.—This is another name for the lapped and mitred dovetail, under which term a description is given. It is also designated secret dovetail.

Mitre Joint Keyed is a mitre strengthened with a slip feather inserted in opposite cross grooves in the butting edges, as in Fig. 122. The same term is applied to the joint in Fig. 123, where saw kerfs are drawn across its plane which are stopped with slips of hard wood coated with glue.

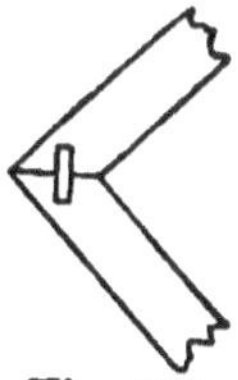

Fig. 122.

Mortise and Tenon Joint.—This differs little from that described in the Carpenters' Section, being usually only smaller and cut and trimmed with more exactness, and sometimes not exceeding in thickness one-fourth that of the framing it secures. Moreover, instead of being fastened with a pin as explained under Pinned Joint in the same section, joiners now generally prefer to wedge up the tenon when inserted in the mortise with one or two small sharp wedges. Double tenons, represented in Fig. 124, usually lie in the same plane, and are advantageous

Fig. 123.

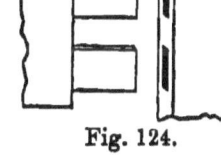

Fig. 124.

in joinery, for consisting in reality of one wide tenon notched out in its central part to prevent bending, it requires less wood to be taken from the piece in making the mortises.

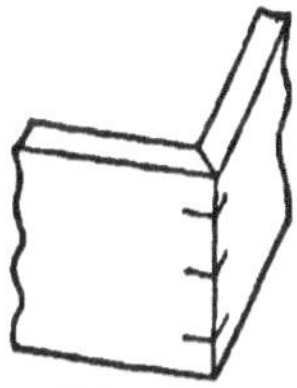

Nailed Joint.—Besides those ordinarily formed, in fixing joiners' work, by the adhesion of the points of nails in wood slips, wood bricks, ranging bond, and specially made unbaked bricks, a good hold may be obtained in mortar joints by using long nails, or by inserting in the joints as the work is carried up doubled-over strips of lead or hoop iron so situated as to seize between the doubling the points of the nails.

Notched Joint.—This is effected in many ways, the common dovetail, for instance, resulting from cutting notches in the edges of two boards in such a way that the notches in the one exactly correspond with the pins left between the notches in the other.

Open Joint occurs when the surfaces designed to be in contact show an intervening crevice. Tongues and rebates in floors, &c., diminish the mischief arising from joints opening by arresting dust and rays of light. To prevent open joints in skirtings, dados, &c., one edge or the middle only of work jointed together, but not framed, is fixed, the other or both, as the case may be, moving in a groove. On the whole, the usual open joints about a well-built but ill-ventilated building are probably more productive of good than harm in gradually replacing vitiated by purer air, and, in fact, acting in the aggregate as an old-fashioned unscientific, silent, and unseen ventilator.

Piecing Joint.—This is a species of scarf, and often occurs in reparations. It usually consists in a glued-up square-edged joint running in one direction parallel to the fibres, but always oblique to them when crossing the grain. When the piecing, however, is merely lengthening, it is effected by some kind of heading joint already described.

Pinned Joint corresponds with that of similar name in Section VI.

Plain or Plane Joint.—This occurs when boards abut against one another with squared edges or sides, as exemplified in a vast quantity of cheap and common work. Even if nailed to joists, rails, ledges, &c., the joints will open, and buckling, twisting, or casting probably display itself at the edges. Perfect seasoning and screw nails are, however, capable of producing a neat junction under such unfavourable conditions. Fig. 125 is an example of a hinged and beaded variety.

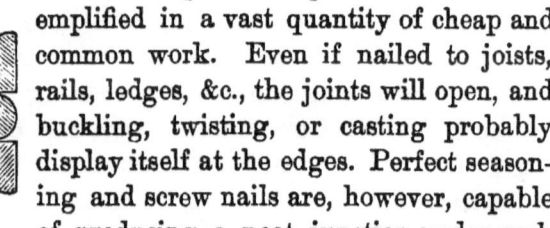
Fig. 125.

Ploughed and Feathered Joint.—This is the same as has been described under the name of Grooved and Feathered, the groove being taken out of the edge of a board by means of a plane called a plough. This tool is not limited to grooving edges only, for it has attached to it in a position parallel to its side, and connected by means of two stems, strigs, or arms passing through mortises in the body of the plane, an adjustable fence which will let the plough off the edge and enable it to make a groove at any distance from it not exceeding the available length of the arms, which is generally 6 in. or thereabouts. The arms are secured by wedges or screws, and the plough usually has a set of 8 irons from $\frac{1}{8}$ in. to $\frac{5}{8}$ in. in width, which can be let down by means of a screw-stop or guard to run a groove as much as $1\frac{1}{4}$ in. in depth.

Ploughed and Tongued Joint is synonymous with grooved and tongued joint and its alternative appellations.

Plugged Joint is the offshoot of plugging, which consists in driving a wood plug into a mortar joint or else wedging it into a hole drilled or cut for it in order that grounds and other joiner's work may be fixed thereto. The cut hole is by far the best, for if carefully made it does not shake the wall nor loosen the joints, and moreover it admits of a larger surface for attachment, which is sometimes desirable, as, for instance, when the smith fixes lever bell pulls. Driving plugs, to hold well, should be cut somewhat crooked so as to grip by resilience as well as by friction, but backings are better secured by spikes and holdfasts, instead of by plugs, for the above-mentioned reasons.

Rabbeted Joint is the same as a rebated joint, under which head it is described.

Radius Joint.—This is synonymous with rule joint.

Rebated Joint is formed by taking a rectangular slip out of the edges of two boards to a depth of half their thickness with a rebate plane or side fillister, or, in other words, by rebating them to that extent on different sides. The remaining projections which are equal are lapped over each other, as in Fig. 126, which brings the surfaces of the boards supposed to be of equal thickness into parallel planes. A rebated joint also occurs when one of the surfaces of contact consists of a rebate of any degree of depth.

Fig. 126.

Rebated and Beaded Joint is a joint between two boards, one of which is rebated, and the other beaded and sometimes rebated as well. Figs. 127 and 128 furnish examples.

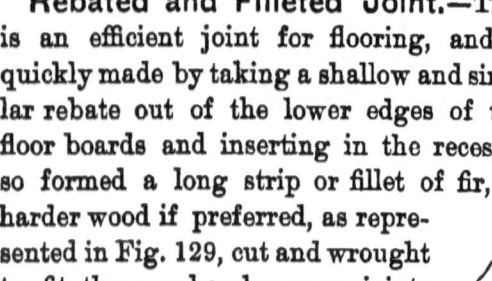

Fig. 127.

Rebated and Filleted Joint.—This is an efficient joint for flooring, and is quickly made by taking a shallow and similar rebate out of the lower edges of the floor boards and inserting in the recesses so formed a long strip or fillet of fir, or harder wood if preferred, as represented in Fig. 129, cut and wrought to fit them, whereby open joints are prevented when shrinkage sets in.

Fig. 128.

Fig. 129.

Rebated and Mitred Joint is a combination of the rebate and mitre, suitable for boards of different thicknesses, the mitre being continued only a short distance within the arris. It is similar to the joint shown in Fig. 119.

Rebated Grooved and Tongued Joint.—This is made by combining a groove and tongue with a rebate, as shown in Fig. 130, but it is perhaps better known under its alternative name of grooved and rebated joint. The opening of the joint resulting from shrinkage is completely covered, and if used for flooring, the nails can be concealed by edge-nailing instead of face-nailing, that is, by driving the nails obliquely through the edges before laying the next board instead of driving them straight through the face into the joist at each crossing, as is commonly done with square-edged boards.

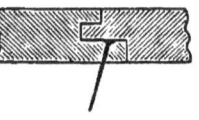

Fig. 130.

Rounded Joint is one formed by taking off the arris, as, for instance, at a mitre, and rounding off the edge to a surface more or less cylindrical.

Rule Joint.—A hinged or movable joint given sometimes to window shutters, but most frequently seen in tables and other productions of the cabinet maker. Fig. 131 represents the joint when open, and Fig. 132 when closed.

Running Joint occurs between surfaces of which one is intended to move or slide on the other, called a runner. Friction, of which there are two sorts, enters into the consideration of this joint.

Fig. 131.

Fig. 132.

One sort resists the commencement of motion, and may often be overcome by a slight jar. The other is a uniformly retarding force, and is proportional to the weight or pressure and independent of the extent of surface in contact. It is, therefore, often inconsiderable between planed surfaces in light pieces of joinery, but when troublesome may be much diminished by rubbing a little dry plumbago on the runner. If this kind of friction exists between metal surfaces, clarified oil should be mixed with the plumbago.

Scarf or Scarfed Joint occurs in the operation of piecing or repairing and replacing the decayed parts of rails, stiles, linings, &c., as previously noticed under Piecing Joint.

Screw Joint is one fixed by wood screws or screw nails, commonly called screws, which are made of brass, copper, or iron. The latter kind are sold in various thicknesses, and in more than a dozen different lengths, ranging from $\frac{1}{2}$ in. to 6 in., many of which are gimlet-pointed or self-boring. They are used in fixing hinges and other metal work, and in situations where the joint may require loosening either for inspection or repair, as for instance at the seats and risers, &c., of water closets, and in beaded casing to piping. There can be no doubt, however, that the joiner does not employ a sufficiency of screws, for many handsome ceilings, &c. have been injured by the jar of the hammer when the screw driver would have harmlessly and more efficaciously performed the work, whilst thousands of rattling windows betray the absence of screws from the joints between the inside beads and linings of the frames. In making a screw joint, the hole in the piece through which the screw first passes must be large enough to admit freely and easily its whole body or stem. The approximate adhesion of a wood screw in timber in lbs. is said to be found by multiplying the diameter, pitch, and length of thread inserted in the timber all in inches, and the product by 42,000 for soft woods, and by double the quantity for hard; but this must be obviously very untrustworthy in the case of the ordinary sized screws used in joiner's work, owing to the great difference in the holding powers of fibrous and vascular layers, and to the many possible positions of the screws with respect to them. Besides which the tensile strength of oak is less than that of fir, whilst there is much doubt as to their comparative powers of resisting a crushing force when both woods are of good quality.

Scribed Joint is formed by scribing, which consists in making a board fit down closely upon an uneven or irregular or moulded surface by marking in the first place upon it with a scribe, or other pointed instrument, the line to which it must be cut. This is effected by placing the board at a little distance from its true or final position, but with one of its finished edges exactly parallel to what will be its own direction therein, and then with a pair of wing compasses, opened to the proper distance, running one point along the profile of the fixed surface whilst the other scribes the required line on the board. Skirtings are thus scribed down to floors and strings much in the same way to stairs, the difference being that a shallow sinking, called a housing, is made in the face of the string after scribing the profile of the step instead of cutting away part of the edge. Similarly, previous to housing the ends of skirtings at a re-entering angle, instead of mitring them, the profile of the moulding has to be scribed, and the joint, though also housed, is often in consequence said

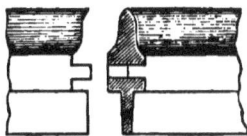

Fig. 133.

to be scribed. Skirtings are likewise scribed to chimney pieces. The way in which the shorter bars of a sash are sometimes cut to fit those which extend through to the frame, as shown in Fig. 133, presents a fair illustration of a scribed joint.

Seam Joint.—This is a plane longitudinal joint.

Secret Dovetail Joint is another name for the mitred dovetail.

Secret Nailed Joint occurs when the heads of nails are concealed from view by an overlapping part. In flooring it is obtained by edge-nailing, that is, by driving the nails in an inclined direction through one and the same edge of the boards only, so as to clear the tongue which confines the edge of the next board, as in Fig. 130.

Shoe Joint is made when the feet of the posts of door frames are fitted with cast iron shoes, having a projecting stub at the bottom for setting and fixing on a stone or wood sill provided with holes for the stubs. A little white lead should be inserted between the shoe and its contents on the one hand and its bed on the other, and the bead on the post should be struck to match that cast on the shoe.

Shutting Joint is usually an open one, varying in its degree of air-space inversely as the fitness of the stuff and quality of labour expended in its formation. In the case of doors, it is situated between the edge of the lock stile and the rebate in the jamb lining or door post, or between the meeting stiles when the leaves fold. In order that the door in superior work may fit close and open easily, the stile or rebate must be bevelled so that the most distant movable point measured in a horizontal direction from the line of axes of the hinges may just clear the nearest point of the jamb in the same horizontal plane. The thickness of the rebates of casement windows is usually too little to need any bevelling, and a little sticking in opening them is not objected to by those who value the exclusion of wind and rain in the winter season. The shutting joints of sash windows are always bevelled except at the top, that at the meeting rails being shown in Fig. 134, and that between the bottom rail and wood sill in Fig. 135, though the rails are not always checked as shown.

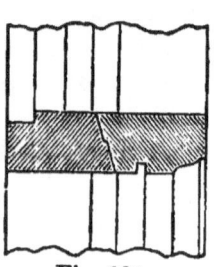
Fig. 134.

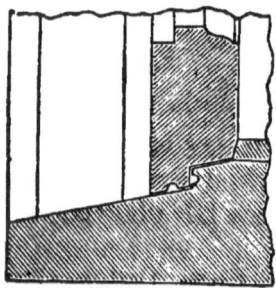
Fig. 135.

Sliding Joint.—In sash windows the joints between the sash stiles and the pulley stiles are sliding ones, as is likewise the case with those at corresponding parts of lifting shutters, &c., and all other kinds of hatches, covers, and screens which open and close vertically within frames. The joints between runners and the bottom rails of doors, shutters, and sashes, &c., which run, roll, or slide with a horizontal movement from the open to the closed position are, perhaps, more appropriately classed as running joints. In external situations care must be taken that proper capping, grooves, and water bars are provided to keep out wet or to quickly carry off from the joint any that finds its way to it.

Slip Joint.—A chase mortise joint, or one formed by slipping the edge of a board, &c., into a dovetailed, rectangular, or other groove or chase, is so called. By cutting channels in the proper direction for the dowel holes, a new board may be slipped into place in a dowelled floor without disturbing the existing dowels.

Slip Feather Joint occurs when equal-sized grooves are worked on the edges of two boards so that there may be closely fitted into and between them either a slip of fir cut more or less obliquely across the grain, or else a slip of wood harder than the boards, or even, as frequently happens, a strip or tongue of hoop iron. This cross tongue or slip of wood is, as well as the iron tongue, denominated a slip feather, which, however, when of wood is sometimes of dovetail section, and then it has to be inserted endwise. The slip feather in a keyed mitre joint loses this designation as soon as fixed.

Splice Joint.—This occurs in handrailing, as noticed under Handrail Joint.

Sprigged Joint is made in joiner's work when mouldings, instead of being stuck or struck in the solid on the edge of framing, &c., are merely worked on separate slips and planted on, or laid in, or bradded, or sprigged, to produce, if possible, an equal effect.

Spring Joint is one that is improperly loose, and often very objectionable in stairs and flooring, causing a springy step or board.

Square Joint.—This is the same as a Butt Joint.

Square Edged Joint occurs in boarding when the butting edges are either merely sawn square or else shot true.

Straight Joint.—One that is continued for some distance in the same direction without a break. In flooring the straight joint floor obtains its name from the longitudinal joints having no breaks, owing to the boards being gauged to the same width.

V-Joint.—This has been described under Chamfered Joint.

Water-tight Joint.—This may be made either to retain or not admit water. A slip feather between boards forms an excellent joint when the former is the object. The angle joint necessary in uniting the parts or sides of water trunks, cisterns, troughs, &c., may be either plane or similar to that known and described as cistern joint, but in any case it must be sufficiently close to be impervious. Cistern covers are sometimes protected at the joints by laying over them saddle-back fillets so throated as to form in conjunction with grooves in the boards water channels along each side of the joint. A junction, however, that is devised not to admit water is, perhaps, more correctly styled a weather or weather-tight joint.

Weather Joint.—This has been already defined in Section III. and elsewhere. Rain and wind are as far as possible kept out of the meeting stiles of casements, as explained under Hook Joint, and at the hanging stiles by opposite grooves in the stiles and rebates forming channels leading to another in the wood sill, from which holes convey the water off to the outside. Sometimes the stiles and rebates are tongued and grooved. It is a more difficult matter to prevent wet from entering between the bottom rails and sill when the casement opens inwards. The most usual device is some form of galvanised iron or metal water bar screwed to the oak sill with an overlapping throated weather board. The water bar may be hinged so as to fall flat when the casement is opened and return to its normal position when it is again closed. A small india-rubber tube fastened to the wood sill by means of a projecting edging or lap, and secured just under the bottom rail, has been substituted for a water bar, and sometimes the rails and pitch pine or oak sill are both rebated and grooved so as to interlock. A grooved channel should always run along the wood sill with communications bored through to the outside to conduct the water that beats into the room back again to its more appropriate quarters. Small fillets, called water fillets, are nailed over the joints of flaps, trap and other doors, &c., to keep out rain, and splash boards are placed for the same reason slantwise on the bottom rails of outer doors. In skylights in a sloping or lean-to roof the frame of the light is grooved to fit down upon a corresponding wood or metal tongue in the curb or lining, and the glass projects over the bottom rail without any horizontal bars. Under Sliding Joint will be found an allusion to a similar protective arrangement suitable, amongst other applications, to a slightly inclined running light of a raised skylight on a flat roof.

Wedged Joint.—The joiner tightens up a mortise joint by fixing the tenon already close home to its shoulders with one or two acutely edged wedges cut out of clean straight-grained wood, coated with white lead ground in linseed oil for external or with glue for internal work. These are inserted and driven in between the tenon and one or both sides of the mortise, which is trimmed slightly dovetail to better take the wedges, and they are found to stop the tenon drawing quite as well as the pins which they have almost superseded. Joiners' work, after being fitted and framed, ought to be left to season as long as possible before wedging up, to do which a dry period is desirable, and care is necessary lest in the operation of cramping and wedging the splitting faculty of the wedge, noticed under this head in the Carpenters' Section, is overlooked and too much strain allowed to risk the soundness of the joint. In housing stairs to strings the joints are usually secured with glued wedges, though some joiners prefer fastening them with nails instead of glue. Cased window frames, when not built in with the work, are fixed in position with single or pairs of wedges driven in at the side between the lining of the frame and jamb recess, and at the head between the frame head at the pulley stiles and the lintel.

SECTION IX.

PLASTERERS' JOINTS.

Angular Joint.—This is formed when the edges of grounds are bevelled off to enable the plasterer upon laying on his first coat of rough stuff to key it well against and behind them. Similarly it is made whenever stucco or plaster is worked to an angular edge by trowelling against cores, or brick cornices, or bracketings, or any other projections on the exterior or interior of a building either when covering it in the one case with ornamental features, or in the other when decorating the walls with dados and pilasters, &c., or groining or panelling the ceiling, or adorning it with coved or other elaborate cornice.

Broken Joint.—When lathing is properly executed the butt joints of the laths are well broken by snatching.

Butt Joint occurs in lathing when the laths are nailed butting end to end and not overlapping, which latter arrangement only prevails in common work.

False Joint occurs in grooving stucco to represent the bold joints of rustic work, or else in simply lining and jointing it to resemble the fine joints of ashlar.

Grooved Joint.—This is formed as a recessed moulding in stucco to imitate rustications. The plaster soffit also between the fascia and wall plate in half-timbered houses, &c., is keyed into grooves run in these pieces.

Key Joint.—Grounds by being grooved or else mitred or splayed, or bevelled, afford a good key to secure the edge of the rendering. Laths are nailed to joists, quarters, battens, bracketings, &c., at intervals apart of about ⅜ in. so that the plaster may pass between or become insinuated, and by curling and widening out at the back, form a key or clench to preserve the junction. The joints of walls are left purposely rough or else raked out with the same object. Rendering and pricking up coats are crossed or scored with a scratcher to give a key to the next coat, and in plastering under fireproof floors, &c., various devices such as dovetail grooves, indentations, and bedded fillets are adopted to yield the necessary hold.

Mitre Joint.—This can scarcely be called a joint since the parts of a plastic substance enter into a somewhat similar union to that effected by fusion. There is some reason, however, in regarding it as one, considering that in running mouldings they are always broken off before reaching the mitre. In rendering walls the plasterer occasionally mitres with the grounds when bevelled to an angle of 45°. In this artificer's work both inside and outside mitres occur at every angular turn or change of direction of skirtings, cornices, mouldings to pedestals, pilaster caps, bases, panels, &c., and it is essential that they should be accurately formed, showing well-defined angles and arrises. They are usually finished off by hand after the longitudinal parts have been worked by running the mould. Where enrichments are used to interior cornices the plaster castings have likewise to be neatly mitred at the angles, bedded and jointed in plaster, and screwed on where necessary.

Tongued Joint is formed when trowelling or working up to grooved grounds.

Toothed Joint.—The joint between plastering and laths is often only a toothed one, because when the surface covered is woodwork, between which and the plaster there is no room for the latter to curl round, the plastering can only adhere by the tooth afforded by the roughness of the wood. In superior work such a contingency is obviated by brandering or counterlathing. When the laths also are nailed with insufficient interval between them, as is by no means unfrequently the case, the plaster mainly sticks by the tooth of the laths. A vertical surface, however, scarcely requires the key or clench between the laths so essential to ceilings and soffits for their security.

SECTION X.

PLUMBERS' JOINTS.

Astragal Joint is a copper-bit joint ornamented with one or more astragals. An astragal is a small moulding of semi-circular profile, and, amongst other uses, was employed by the ancients to mark the division between the capital and shaft of a column, but it is questionable to which of these members it really belongs. Its position, however, there and elsewhere, sometimes gives the idea that it was used to conceal a joint, and plumbers appear to have wisely adopted it for a somewhat similar purpose. When attached to round pipes astragals are formed either by dividing a small pipe longitudinally into halves, or else by casting them in moulds, and in either case they are bent round the enlarged or female end of the pipe, and neatly soldered on with the copper-bit and fine solder. When the socket happens to be cast, it is usual to cast it complete with astragals and ears.

Autogenous Soldered Joint is made without solder by merely scraping or cleaning, and fusing by means of a blowpipe and a mixture of hydrogen gas and air, or other appropriate gas, the edges to be united, which are placed sufficiently close to run together, otherwise a stick of the metal must be fused at the same time and allowed to drop upon them. It is useless in plumbers' roof work, because the intense heat required to melt lead would be followed by an amount of coolness and contraction that would inevitably produce cracks in proximity to the joint, having regard to the friction between the lead and the underlying surface.

Bed or Bedding Joint.—Great care is necessary in bedding w.c. trunks, basins, soil-pipes, traps, &c., so that there may be no settlement to cause lower parts to receive support from upper ones, and then when the strain is too great to break or draw away from them at weak sections or imperfect joints. Though not falling within the plumbers' domain, it may be here observed that in bedding soil pipes, in that part of their course from the inside to the outside of a building which passes through the wall, it is essential that their position should be such as to enable the joints to be reached at every part, and where stoneware pipes are used the same precaution is necessary throughout their whole downward course. Where the drain-pipe is not ventilated, as is unfortunately too frequently the case, it is obvious that no joint can be pronounced secure against the passage of mephitic vapour unless absolutely and unalterably non-porous, and certainly neither clay nor cement can be so styled. In some instances lead soil-pipe is joined on to a stoneware socket system, in which case the junction, after being made with yarn and clay in the usual manner, may be top coated with a mixture of resin and tallow, described under Cement Joint, to the depth of half an inch or thereabouts.

Block Joint is a mode of connecting lengths of vertical lead soil or waste pipes on wood blocks about 10 feet apart, and which are usually built or let into chases left in the walls for the pipes. A lead flange about 5 in. larger than the diameter of the pipe, and with a central orifice a trifle larger than the pipe, is worked round a hole cut and dished in the block, as shown in Fig. 136. The lower

Fig. 136.

length of the pipe is then passed up through the hole so

that its funnel may be tafted back upon the flange, after which the top length is prepared by shaving, &c., and stepped into the funnel, and the whole are soldered together by wiping, as described under Wiped Joint.

Blown Joint is a soldered joint made with a blowpipe (under which name in Section XIV. it is more fully described) or æolipile, otherwise blowing apparatus or blowing lamp. When this joint occurs in plumbing it is not wiped, though, on the other hand, a wiped joint may be made with the above instruments and strip wiping solder. In forming the joint between two pieces of lead pipe, the end of one piece is rasped off at the edge and inserted into the end of the other previously tafted with the turnpin, tanpin, or tompion, as the tool is indifferently called, and the cup is filled with fine solder, described under Soldered Joint, and in the manner explained under Blowpipe Joint. In any case, whenever the elements to be joined are lead ones, whatever may be the form of the edges, the parts where the solder is to stick, and of course unite, must be brightened with the shavehook, well dusted with resin—or covered with whatever flux may be used—and floated with solder so as to get the work into combination with it, and not merely in superficial contact, the strip solder being dipped by some artificers into powdered resin, and held to the flame, whilst the excess is removed, if necessary, after becoming hard by filing, &c. Brass and iron fittings must be previously tinned, as noticed under Copper-bit Joint, and smudge or soil, described under Wiped Joint, may be used if desired to prevent the solder from sticking beyond the confines of the joint.

Bottle Nose Drip Joint occurs occasionally, and is, as its name implies, a variety of drip possessing somewhat more character than the common sort. The rough boarding terminating the drip step is finished off

Fig. 137.

with a nosing, either square or round, called in this instance a bottle nose, and the lower sheet of lead is dressed down so that its upstand stops against it, as in Fig. 137. The upper sheet is then dressed round and over the nosing and upstand, but not quite reaching, for the reasons given under Drip Joint, the nearly horizontal surface of the lower level of the drip.

Branch Joint is one occurring between a main and branch. Sometimes the latter starts from the main at right angles and at others obliquely, but in both cases its end is first prepared by rasping, and the size of the necessary incision struck off from it. The hole is then made with a gimlet or red-hot iron, and enlarged with a bolt or other tool so as to leave the edge slightly upraised. The two pipes are then fitted without allowing any burr to remain in the bore, and after being prepared and put together in the usual way for soldering, are ultimately dexterously united by wiping.

Burnt-in Joint.—The lead of gutters is secured to grooves in stone cornices, and that of flats to sills, by turning down the edges of the metal into the grooves and pouring nearly red-hot lead upon them until the whole become blended into one mass. Should moisture be present, a little resin spread along the groove before pouring will prevent the molten metal flying out in spray. The joint is finally caulked when cold. Similarly a lead pipe is joined to another by pouring lead round the junction instead of solder. The lower of the two pipes is first opened out and then stopped up with sand nearly to the throat of the opening, a wad being placed on the top of the sand. A funnel is then attached outside, and the tafted edge of the pipe surrounded with a small upstand of sheet lead, the funnel being fitted with sand up to the level of its top. As soon as the upper length is inserted, having been previously rasped and fitted, and both pipes being independently secured, the red-hot metal is poured into the cup so as to fill it without leaving vacuities or air holes. In other cases the edges of halves of pipes, traps, &c., are united by melting off lead from a stick with a red-hot iron.

Butt Joint is made between two pipes by means of a steel core or cast iron cylinder fitting the bore of the pipes when raised to the same heat as the solder, and attached to a chain for regulating its position and extracting it. The ends of the pipes are cut and shaved truly to a close butt and butted together over the cylinder, the joint being wiped in the usual way and finished off with the common bulbous protuberance.

Caulked Joint.—This occurs when molten lead is used to form a joint between stone and ironwork, or as noticed under Burnt-in Joint and Run Joint. As the metal contracts in cooling, it has to be "set up" or caulked, that is, punched into its seat when cold.

Cement Joint.—A mixture of tallow and resin is denominated plumbers' cement, being useful amongst other purposes for securing the bottom valve of a pump. The two ingredients are carefully melted and mixed over the fire in such proportions as to make the mixture sticky and stringy, and not greasy or brittle. Sometimes brick dust is added. In this trade cement is a general term for any kind of stopping used for joints. It includes putty, white or red lead, or the two combined (described under Flange Joint), &c., all of which, however, being liable to be cracked by shaking or settlement, or to become loosened by drying up, can never be wholly depended upon for the total exclusion of sewer gas unless used between flanges or screw threads.

Cistern Joint is a wiped soldered joint, the water preventing tension.

Cone Joint.—This is a plan adopted by Messrs. T.

Lambert & Sons to connect lead pipe to wrought iron tubing without solder, and is shown in Fig. 138. A gland nut having been passed over the lead pipe, it is opened out by means of a turnpin having the same taper as the purposely tapered end of the iron main or wrought iron tubing, so that the two ends only have to be drawn together by bolts and nuts to make a perfectly sound joint, whilst the fullway is preserved.

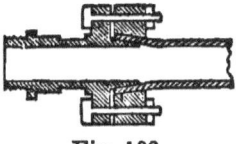

Fig. 138.

Copper-bit Joint is a soldered joint made with a copper-bit and fine solder in strips, sticks, or cakes, the latter of which are about 4 in. by 6 in., and from ¼ in. to ⅜ in. thick. Sometimes, however, the position of a joint is such that it is more convenient to use the bit than the bulbous grosing iron after the molten solder has been poured over it, but this of course is not a genuine copper-bit joint. The copper bit, called also a soldering tool or soldering iron, two varieties of which are shown in Figs. 139 and 140, consists of a bit or bolt made of copper, having a hatchet-shaped or else a pointed nose or end. It is held by a wooden handle secured to a short iron shank riveted to the bit. The hatchet-shaped variety is generally

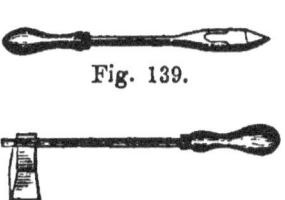

Fig. 139.

Fig. 140.

used by plumbers, who often call it from its shape the hatchet bolt, whilst gasfitters find the pointed kind most convenient. Before using the bit it has to be tinned, that is, a face of solder is given to it, which is effected in various ways, an old fashioned one consisting in heating it to a dull red, then cleaning it by quickly filing bright its nose or extremity, rubbing it instanter upon a piece of sal-ammoniac (muriate of ammonia), and immediately afterwards on a "tin pan," which is a piece of tin or copper plate on which a few drops of solder have been deposited. Finally, if necessary, it is wiped with a piece of dry tow or cloth. It can then be re-heated and used at a great heat, but not made red-hot, else the face will be burnt off. The more modern plan is to use plenty of spirit of salt (hydrochloric acid) killed, that is, turned into chloride of zinc by putting scraps of zinc into it. When ebullition ceases it is diluted with a little water, and after cleaning as above, the hot end of the tool is momentarily plunged into it and then touched with solder, and similarly touched and plunged again and again until the face is nicely tinned. The bit must always be tinned with the same flux as is used for the soldering, so that if resin is used the extremity of the bit prepared as already explained has merely to be well and expeditiously rubbed upon a mixture of a little powdered black resin and a few drops of solder. In soldering together the edges of zinc, a face having been put on the bit with killed spirit, this fluid is applied to the work with a small brush or piece of stick or cane, and then the strip of solder is drawn along the joint with the left hand whilst the copper-bit follows it in the right. Care must be taken to wipe off all stains of the salt with a wet cloth. Pieces of lead are somewhat similarly united with resin as a flux. Brass and wrought and cast iron piping and fittings are brushed with the above liquid, or a solution of sal ammoniac, or unkilled hydrochloric acid, which is the best for iron (the iron surfaces being first filed clean and bright), and then thoroughly well tinned with a little of the solder and tinned copper-bit, otherwise the solder will not hold properly. The joint is then made by expeditiously floating more solder and well sweating it about the parts. Cast iron requires a very hot bit to make a good joint, but of course not hot enough to take its face off, and the iron must be scraped or by some means or other got quite bright before tinning, which if perfectly well done as just described will give the solder an inseparable hold.

Dog's Ear Joint.—This is formed by turning up two adjacent edges of a piece of sheet lead at right angles, pinching the corners together, and then doubling back the pinched-up part against the upstand, as in Fig. 141. It is a joint used for dressing up lead against the end of a gutter, &c., instead of bossing it up, and making it is sometimes denominated pig's lugging instead of dog's earring, but even to a connoisseur in such matters there can be but little choice between the two appellations.

Fig. 141.

Double Cone Joint is a mode of connecting two lead pipes without solder and very similar to the cone joint already described, being manufactured by the same well-known firm, and intended for situations where skilled labour is difficult to obtain. It is made by opening out both ends to be joined with a tanpin of the same degree of taper as that of the two ends of a loose piece of brass having a corresponding bore to that of the pipes, and which can thus be inserted into their enlarged ends. Previous to their enlargement, however, a gland nut is slipped over each end, so that all that remains to be done to make a sound joint is to draw the nuts together with bolts in a manner exactly similar to that represented in Fig. 138.

Drawn Joint results from the process of drawing or making a seam by floating the solder along it with the ladle and iron.

Drip Joint is represented in Fig. 142, and is a means adopted for joining sheets of (7 lb.) lead in gutters and flats to allow for contraction and expansion. When the joint is across the current the plumbers' drip occurs about every 9 ft. or so, and consists of a low vertical step in the boarding, from 2 in. to 3 in.

Fig. 142.

deep, having a rebate to receive the lower sheet, which is dressed close down over it. The upper sheet is then turned down over the upstand and dressed to within 1 in. or ¾ in. of the surface, in order that dirt may not collect at the bend and cause damp to rise by capillary attraction. When the joint is in the same direction as the current it can be formed as shown in Fig. 143. The bottle nose variety has been already explained.

Fig. 143.

Elastic Joint.—This is noticed under Expansion Joint.

Elbow Joint is one formed at the end of an elbow which may be either screwed for iron with or without a running nut at one end, whilst the other is tinned for lead or else provided with a union similarly tinned. Or the elbow joint may be of the kind described under Cone Joint. In lead piping elbows were formerly made instead of bends by cutting one pipe and inserting the other at the mitre, resulting in an upstand being left inside the pipe from too much overlap, besides impediments in the shape of protruding spurs of solder. Elbow joints are now made by cutting a piece out of the throat of the pipe, pulling it together to the proper bend, and working the lead neatly over the throat so that there may be no obstacle left to the flow, and finally soldering it up with a neat tight joint, which need not extend all round to the heel.

Expansion Joint is one devised for averting the disastrous effects of interrupted expansion and contraction due to changes of temperature. The mediæval plumbers as a rule laid lead as loose as possible, and were as lavish of lap joints as sparing in the use of solder. In metallic roof coverings rolls and drips and flashings are the usual artifices. In order to render their joints air or water tight, and at the same time to allow longitudinal motion, cast iron socket pipes have their cavities between the spigot and faucet closely packed with gaskets and grease, or hemp and red lead, or rings of india-rubber in the case of hot-water pipes, as described under India-rubber Joint. An expansion of 1 in. is usually allowed for every 80 feet of cast iron piping. Long lengths of ventilating pipes require expansion joints which, when of lead, are frequently made as described under Slip Socket Joint. An elastic joint is merely another term for an expansion joint invested with slight powers of extension through the medium of an elastic substance, which cleaves sufficiently tenaciously to the parts connected during their slight motion to preserve the tightness of the junction. Amongst the patent varieties Messenger's is a neat and economical example. This is intended for hot and cold water and gas, and consists of an india-rubber or other packing retained in the cavity of the socket by means of a loose iron ring encompassing the spigot end, and which is drawn up and bolted to the socket with nuts and screw bolts.

Faucet Joint.—This may be either a socket, slip socket, or astragal joint.

Flange Joint is a wiped soldered joint between two pipes, and is made with one flange only, which may be tafted back by degrees and by taking great care after the pipe has been opened out, but some plumbers cut the flange out of a piece of milled lead with the necessary orifice and to the desired diameter. This detached flange is then placed over the hole in the floor or cistern, &c., where the joint is to be situated, and the pipe is passed up through it without the edge being rounded off as with block joints. Its end is opened out with the tanpin as described under Taft Joint, and the flange and pipes being all properly shaved and prepared for wiping, the whole are united in the usual way quickly and with plenty of solder. Another kind of flange joint occurs when a lead flange is soldered to a lead pipe to correspond with the flange of an iron branch to which it is to be joined.

K

Red lead and hemp packing are inserted between the flanges, and an iron ring or washer is placed behind the lead one previous to bolting all through together with screw bolts and nuts.

Flashed Joint.—A kind of lap joint formed by covering with a flashing the upstanding edge of a sheet of lead in a gutter, &c., or protecting by the same means any part of a roof cut into by the protrusion of gables, dormers, chimneys, skylights, &c. A flashing consists of a strip of 5 lb. lead, from 5 in. to 16 in. in width, secured along one edge to the wall with wall hooks, or else by tucking it into a groove or joint, whilst the other edge is left free and dressed so as to fall about 3 in. over the upstanding edge, which is about the same width as the flashing, or over or under slates, &c., as the case may be, so as to make all weather-tight and avoid checking any lateral motion of the lead during changes of temperature. In those cases where the flashing is notched and let into a succession of joints one above the other to accommodate it to an inclined upstand, it is termed a stepped flashing. The flashing at the feet of chimneys tucked into a joint, and lying down on the roof so as to allow the side or raking flashings to lap over it, is termed an apron or apron flashing, which title is applicable to any flashing that hangs over and covers an upstanding edge; but where an apron occurs, the flashing under it is not fixed.

Flexible Joint is formed between two pipes with plain ends united with a piece of india-rubber tubing.

Float or Flow Joint.—This is the ordinary copper-bit joint already described.

Flush Soldered Joint occurs in lead flats, &c., where it is found necessary and safe to unite sheets without providing for expansion. It is made by sinking a chase or groove from $\frac{3}{4}$ in. to 1 in. wide and about $\frac{1}{2}$ in. to $\frac{3}{4}$ in. deep in the boarding, according to its thickness and that of the lead, and bending the edges of the sheets into it, allowing a little overlap, and nailing them down to the chase with clout nails. The usual scraping, soiling, &c., having been performed, solder is then poured into the groove over the edges and floated and smoothed with the iron, and all wiped smooth and flush.

Heading Joint.—This term is used for a lap joint crossing the direction of the current.

Hollow Roll Joint is used on flats in the direction of the fall, and is made by turning up against one another the edges of two sheets of lead, allowing one to stand a little higher than the other so as to be bent over it, and then rolling the two over together, as shown in Fig. 144, with the tingles or latchets between them. Wood rolls are thus dispensed with.

Fig. 144.

Knee Joint is a term used sometimes for elbow joint.

Lap or Lapped Joint.—This is made when from 3 in. to 6 in. of one sheet or strip are laid over the upstanding edge of another, as in fixing aprons or flashings. It likewise occurs when the edges of lead roof coverings are dressed flat over those of others, as in heading joints or in the corresponding junctions which are formed when more than one length or strip are required in covering hips, ridges, and valleys, with the same metal. In ancient times, sheets of lead, when circumstances would allow, were lapped over the ridge of a roof so as to hang some little way down on the opposite side.

Lead or Leaded Joint.—There are three varieties of this joint for which the plumber provides the lead, but he rarely makes them, as this artificer usually employs solder or cement for jointing, excepting when burning in with molten lead. One variety is run with molten lead, another is made by inserting into the pipe socket or slipping over the spigot a ring of cast lead and then forcing the two ends tightly together, whilst the remaining one consists in placing an intervening plate of lead between bearing surfaces to obtain a fine joint, equally distributed pressure, exemption from damp, or a conjunction of one or more of these desiderata.

Long Joint.—A round wiped soldered pipe joint, in which the length of the egg-shaped protuberance or joint is about one and a quarter times its breadth.

Longitudinal Joint.—This is made when two edges are united, either on the roof, &c., or in pipe-making. The weld or seam of the pipe made by hydraulic pressure and not drawn is said to be sometimes so defective as to yield to very slight pressure. The best pipe is made by pressing the molten metal between a tube and mandril by hydraulic pressure, and not by drawing. This sort is jointless. Another kind is noticed under Seam Joint.

Mitre Joint.—An obsolete method of forming a branch joint by cutting out an angular notch in the main pipe and a corresponding projection in the branch and mitring the two with solder. However well done, it is wasteful of solder and wholly unfitted for soil-pipes, being both weak and altogether wanting in the proper smoothness of interior required for clean flushing.

Overcast Joint is another name for striped joint.

Overcast Ribbon Joint.—This is a ribbon joint made as explained under that head, but finished off by being striped or overcast with the copper-bit in such a way as to present, instead of a smooth surface, one covered with facets.

Overlapping Joint is a mode of connecting two pieces of lead in the direction of the fall or current sometimes, but not by any means frequently, adopted, and is made by turning up two contiguous edges unequally, one about $1\frac{1}{4}$ in. and the other about $1\frac{3}{4}$ in., so that the additional half-inch may be carefully turned down again over the $1\frac{1}{4}$ in.

upstand. This being done the two edges are well and closely dressed together with or without tingles between them, and are either left standing up or folded and doubled and dressed down close to the surface, as in Fig. 145. No roll is used, nor is a hollow left in its interior as in the hollow roll joint. This variety thus compactly rolled together is otherwise termed a welded joint, and is very suitable for balcony and verandah floors, &c.

Fig. 145.

Pipe Joint.—Lead and compo. tubing are joined by means of cone joints, or ordinary unions soldered on, or else with wiped, blown, or copper-bit joints. Different lengths of lead soil and waste pipes are united with flange, block, wiped, copper-bit, or slip socket joints, the latter not being applicable to soil-pipes, which in common work are often lengthened by working one end a little way into the other and soldering them together with the copper-bit. When the work is properly done vertical soil-pipes have tacks of 7 lb. or 8 lb. lead every 3 ft. or 4 ft. apart, not less than 10 in. by 9 in., strongly soldered to pipe so as to secure it to the wall on alternate sides with about three wall hooks, or if there are flange or block joints one tack between every two of them suffices. Waste pipes must be protected from breaking away at the joints by similar precautions, though the tacks may be proportionately smaller according to the weight it is their province to support. Pipes of large diameter are usually made as noticed under Seam Joint.

Plugged Joint.—This is effected by drilling a hole with a stone drill and working it larger at the back to prevent the plug from drawing. The proper-sized screws selected for fixing the work are then carefully covered with black lead and dusted over with the same in the state of dry powder; and their heads being set in a piece of plastic clay in such a way that the stems enter the hole, the clay is worked round the outside to form a luting with an opening at the top. A ladleful of molten lead is then taken and run into the hole, and when set the clay is removed, the lead chipped off flush with the face, and the screws withdrawn with a screwdriver, leaving perfect threads in the plug by which they can be driven in again at pleasure with the work or fitting attached. Wood frames are thus often secured to stone reveals and jambs.

Putty Joint is made in connecting a lead service pipe with a water-closet basin. The end of the pipe is cut to fit the arm of the basin, and after painting it and the inside of the arm as well, the pipe is inserted and the joint carefully stopped and surrounded with red cement, described under Flange Joint in Section VII. A strip of canvas, about 3 in. wide and 3 ft. long, is then saturated with paint and bound round over the joint so as to completely and tightly envelop it, and this is secured in place by winding closely around it from one end of the joint to the other a piece of stout string properly secured and fastened off at both ends, the whole being painted over and left undisturbed to harden. The joint between the basin of a pan closet and the apparatus is made with common glazier's putty, but in the case of a Bramah red lead is used.

Raglet Joint.—This is formed by securing in a raglet, described in Section III., the edge of a piece of lead by means of lead wedges or bats placed about 8 in. apart and tightly driven in. A stopping and pointing of cement or mastic finishes the joint by filling up the groove. Sometimes when the raglet is situated on the top of a blocking course or otherwise favourably for the operation, molten lead is run into the groove instead, in which case the joint is rendered tighter by setting up the lead when cold with a caulking iron to counteract the effects of its contraction. When treated in this manner the raglet must be cut rather wider, and the apron or flashing is said to be burnt in.

PLUMBERS' JOINTS.   199

Ribbon Joint is a copper-bit joint made with fine solder so disposed and left round the junction of the two pieces as to resemble a smooth ring about 1 in. wide and $\frac{3}{16}$ in. thick.

Roll Joint is made when two sheets of lead are connected on domes, or at the intersection of roof planes, or on a terrace or flat, &c., in the direction of the fall or slope. A wood roll or core, about 2 in. square, with its upper surface planed and rounded off and lower edges chamfered, is nailed on the boarding or otherwise secured to the

Fig. 146.

surface to be covered, after which the two sheets are rolled out and dressed flat, their edges planed straight, and then one edge is dressed closely down on the roll reaching to its top, and the other dressed and close hammered again upon that, so as to be outside and surround the roll three-quarters of its girth, as in Fig. 146. The turning down is so arranged that the outside edge is turned away from the most exposed quarter. This joint is impervious to water so long as it does not reach the top of the roll. In other cases the roll is made without the wood core, the edges of the lead being planed straight and set up against one another, the upstands being unequal, or about $2\frac{1}{2}$ in. and $3\frac{1}{2}$ in. respectively. The highest edge is first bent down over the lowest, and then both are turned down together, as in Fig. 147, so as to form a hollow roll. In order to keep the roll down in its place, clips, latchets, or tingles, which are pieces of lead about

Fig. 147.

5 in. by 3 in., are nailed to the boarding at intervals of 3 ft. or thereabouts in sinkings made for the purpose, so that the ends nailed down are flush with the surface whilst the other ends stand up between the set-up edges of the sheets so as to admit of being turned over and down with them in form-

ing the roll, as shown in Fig. 144, which represents the same joint in another of its many slightly varying forms.

Rolled Joint.—A wiped soldered joint between a pipe and union, effected by rolling the pipe, with the union temporarily attached to it with small wedges or any other equally effective contrivance, whilst the cloth or solder is held under the junction with the right hand.

Round Joint.—This is the common egg-shaped wiped pipe joint, as shown in Fig. 150.

Run Joint.—The form of joint about to be described is, perhaps, most frequently made by main layers, but it is sufficiently appropriate to be included in this section. It is a lead joint in piping between two cast iron socket pipes, and is made by inserting a spigot end into the socket of an already laid pipe and caulking the cavity between them with tarred gaskets of spun yarn, or else with white spun yarn if for water, until it is filled to about half its depth. Molten lead is then poured in, which requires great care so as to insure the joint being full enough but not too full, for in setting up there is almost equal danger of fracturing the socket if too much or too little lead be used, the caulking iron in the one case being perhaps applied too vigorously, and in the other running the chance of being driven too far in under the socket. A belt of plastic well-tempered clay is first passed round the pipe close to the edge of the socket, working it up from the bottom, leaving enough room for the lead and an opening or gate for the molten metal to enter. When the lead is cold the excrescences are cut off with a chisel, and then with a steel "set," or stout cranked caulking iron and heavy square faced hammer, the workman sets up the lead, by which means it is tightly wedged, all pin holes arising from air bubbles closed up, and the joint rendered air-tight. In some instances when large pipes have to be united the luting or band of clay is superseded by a flat metal ring, about $\frac{1}{4}$ in.

thick and 2 in. wide, made in two pieces to suit the diameter of the pipe and with the necessary gate. It is hinged in the middle and flanged for bolting at the free ends. This is passed round the pipe close to the socket, bolted, and made tight with clay, and is then ready for pouring in the lead. This contrivance admits of ready application and quick removal for the next joint, and a similar treble jointed iron collar is sometimes used for small pipes. To prevent the dampness of the clay, &c., blowing out some of the lead in spray through the vapour caused by the hot metal it is advisable to drop a small quantity of resin through the gate. Salt and grease are said to have a similar effect. In every case the ladles should follow one another sufficiently quickly or be large enough to hold more than the joint requires, so that there may be no risk of not filling it at one running. Pipes laid and united in this manner can be deviated considerably from a straight line, but a sound and durable joint cannot be made if the spigot be much smaller than the socket. To give an idea of the dimensions of the joint and of the trifling variation in its size caused by the increased diameter of the pipes, it may be stated that for pipes of the respective diameters of $1\frac{1}{2}$ in. and 12 in. inside measurement, the corresponding depths of socket are 3 in. and $4\frac{1}{2}$ in., whilst the thickness of the lead joint is respectively $\frac{1}{4}$ in. and $\frac{3}{8}$ in., and its depth $1\frac{1}{2}$ in. and $2\frac{3}{4}$ in. Thus, although the diameter of one pipe is eight times that of the other, the thickness of the lead joint is only half as much again, and its depth not double that of the smaller one, whilst the depth left in the socket for the caulking of spun yarn is about $1\frac{3}{4}$ in. for both and all intermediate sizes. All lead joints between iron pipes are very easily ruptured unless the pipes are bedded so as to be uninfluenced by superincumbent pressure and protected from shocks. The beds of drums or frusta of stone pillars already plumbed true and kept in proper position by wedges of lead, iron, &c., were occasionally in mediæval times run with lead to a thickness in some places of ¼ in., the joints being very rough tooled and the lead affording the requisite uniform bearing. There are reasons for supposing that the practice may have arisen from a wish to avoid the delay that waiting for the setting of the mortar would cause. However this may be, it is not even now quite obsolete with respect to tracery and the shafts of pillars, though there are several existing instances where splits and fissures owe their origin to the holes through which the lead was poured. Raglets, cramp holes, holes for lewis bolts, holes in hook stones for crooks, and in sills for guard bars, mortise holes for railings, lead plugs for screws, &c., are run in the same manner, the clay luting being in each case round or under the hole, or placed to the best advantage for conducting the molten metal all over the joint. Care is always necessary that iron embedded in stonework is completely surrounded so that moisture cannot reach it, otherwise the formation of hydrous oxide will certainly corrode, exfoliate, and swell it, and burst or blow the work. Cast iron resists oxidation perhaps much better than wrought if not denuded of its so-called skin (p. 145), but so far as concerns the efficacy of lead as a protective coating, its value is diminished if by any means the lime in the stone or mortar succeeds in oxidising the lead and dissolving the oxide.

Screw Joint is used in attaching lead covering to boarding, &c., with wood screws. Sinkings are made in the wood where the screws are to go, the lead is dressed down into them, and one or more screws are driven home. Their heads consequently lie in small hollows, and these are filled with molten solder wiped flush with the general surface. The patches of solder are called lead dots, and preserve the screw heads from oxidation.

Seam Joint is a mode of uniting the edges of sheets of

metal by means of a seam with or without a lap or fold. In the latter case solder is necessary. A seam joint between two sheets of lead without a roll is formed by turning up the adjacent edges unequally so as to admit of one being bent down over the other. When this is done they are both turned or folded and dressed down close together flat to the surface and the joint is completed, but it is not equal to the roll joint either in neatness or security. When the plumber forms pipes of large diameter out of sheet lead the seam is soldered either with a copper-bit or red-hot iron. In both cases the edges are planed true, the lead rolled round a wood core, and the edges brought together and soiled, shaved, and "touched" or greased with tallow to prevent retarnishing. If the seam is to be made with the copper-bit, it is done with or without soil, and according to the usual method adopted when using that tool, as already described. When, however, it is made with molten solder and the iron, the solder is poured on with the right hand so as to cover about a foot in length and left floating, whilst with the left hand the plumber draws the red-hot iron along each side of the seam and cuts off the surplus solder. He then proceeds with another short length in the same way, an assistant in the meantime pouring or swabbing with a sponge a little water on the finished part to cool the solder and keep it from opening. This operation is sometimes called drawing lead pipe, but it is not to be confounded with that of making pipe by pressing or forcing lead through an orifice provided with a central mandril by means of a hydraulic press, nor with the older method of drawing cast lead pipes through a series of openings in steel plates to reduce them to the required thickness, and from which the term drawn pipe has originated.

Short Joint.—A round wiped soldered pipe joint of which the length is rather less than the breadth.

Slip Joint is made by inserting the end of a lead pipe

into an iron soil-pipe, for instance, of larger bore with a tight packing of red lead and hemp between them, and a stout india-rubber band, 3 in. or 4 in. broad, surrounding the joint on the outside. This joint also admits of being stopped with molten lead or iron cement, but it then ceases to be movable. Slip joints undefended by traps have been fruitful in causing ill-health, if not mortal sickness, but when traps exclude sewer gas no pernicious effects appear to be due to them, for such joints occur between the flanged necking of a w.c. apparatus and the D trap below it, the pipe of the trap being dressed back upon the floor, or otherwise as considered expedient.

Slip Socket Joint.—This is a socket joint furnished with a contrivance, as shown in Fig. 148, to allow a length of metal piping freedom to expand and contract during the passage through it in almost immediate succession of hot and cold water. In the case of lead pipes a flange is soldered to the foot of the upper length, and an india-rubber ring inserted between it and the flange of the socket, with or without a packing of red lead and hemp in the cavity round the spigot. It must on no account be attached to inside or even to any unventilated soil-pipes, which when fixed outside a building should if possible run down a shady corner or other sheltered part to escape the ill effects of too great a range of temperature.

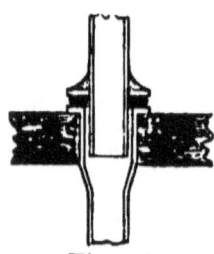

Fig. 148.

Socket Joint.—This occurs in lead rainwater pipes, vertical waste pipes, &c., the sockets being usually cast with ears and mouldings if required, and soldered on with the copper-bit. Generally, however, in lieu of a socket the tacks and astragals are soldered to the upper end of the lower pipe, and the lower end of the top pipe is reduced and inserted therein.

Soldered Joint is made when two metal pipes, sheets, parts, or pieces are united by means of a third substance called solder, which is a metal or alloy varying in kind and chosen for being more fusible than the metals united, and capable of combining readily with both. The soldered joints made by builders belong to the trades of the plumber, zincworker, and gasfitter, and the solders employed by them are mixtures of lead and tin, and called "soft" solders, being so named because they melt at low temperatures, whilst the "hard" solders (described under Brazed Joint in Section XII., and used by the smith and coppersmith) melt only at a red heat. The soft solders are again divided into *coarse* and *fine*, the least fusible containing the most lead or the least tin being classed as coarse, whilst the most fusible or those containing a preponderance of tin are called fine solders. As lead and zinc when used as a roof-covering move under the influence of a change of temperature, no soldered joints are proper between sheets, though in minor parts, or in patching or effecting repairs, they are permissible. If one edge also of a sheet be left free the other may be soldered, but the necessity for this rarely arises, since the fixed edge is usually secured to a raglet or joint in masonry, or by means of "lead dots," and screws, if to woodwork, as noticed under Screw Joint, though nails with large heads may be used instead of screws. The lead lining to stone gutters is sometimes jointed together in lengths with solder, but the whole is left free. So good, however, was the masonry of the gutters in mediæval buildings that no lead lining to them was necessary. The parts to which the solder is to cling are scraped clean with the shavehook and touched with tallow to prevent oxidation, and smeared with tallow as a flux, and are then run with the usual coarse and soft solder called pot metal or plumbers' solder, composed of 2 parts of lead and 1 of tin, which is smoothed down with the red-hot iron and finished off by wiping, as more fully described under Wiped Joint. The connection between the outgo of lead traps and soil pipes ought invariably to be a wiped soldered one. With fine solder consisting of 2 parts of tin to 1 part of lead, or these metals in slightly different proportions, resin is the flux used. The fusing points of lead, zinc, tin, and these varieties of coarse and fine solder are 617°, 773°, 451°, 446°, and 340° Fahr. respectively.

Striped Joint.—An old-fashioned variety of wiped pipe joint, characterized by having the bulbous protuberance ridged or ribbed and somewhat unfinished in appearance when contrasted with the smooth symmetrical and turned-like surface of the ordinary wiped joint. The striped variety is said to be stronger than the other owing to the solder being less disturbed in setting, which is owing to the plumber using less solder, and the wiping cloth less, but the iron more. Others attribute its greater strength to the process consolidating the solder and preventing "weeping," whilst others again maintain that its only advantage is to be found when very coarse solder—3 or 4 of lead to 1 of tin—is used, which would set quickly and be porous were it not glazed over by striping or overcasting. The ribs, ridges, or furrows are made with the heel of the iron when barely hot enough to melt the solder, but whilst the solder is still hot, by passing it up and down all round the joint from end to end, thus leaving a series of lateral flutes about $\frac{1}{2}$ in. wide at the centre and dwindling away towards the extremities.

Swivel Joint is the same as described in Section XIV.

T-Joint.—This can be made like the common branch soldered joint, or a T may be inserted as explained under Union Joint, but this plan in water-fittings is not considered so suitable as a well-made branch joint.

Taft Joint is used for uniting lead pipes, and is made

by tafting or slightly enlarging or opening the end of one pipe and inserting therein to the depth of about ¼ in. or ⅜ in. the lightly rasped end of the other, which must nicely fit it without any burr being left on the inside, or in any way interfering with the full bore. The lead round about the joint must then be shaved perfectly clean and bright, the ends brought closely together again, a little resin shook around the junction, and the cup filled with fine solder with the blowing lamp. As a general rule the edge of the inner pipe should not face the current. Another kind of taft joint is made when lead pipes are soldered to cesspools or gutters, surrounded on all sides by walls or parapets to conduct the rain to ordinary down-pipes or rain-water pipes. Before laying the lead a hole is cut in the gutter boarding sufficiently large for the pipe, and sufficiently sunk to allow the gutter lead and the tafted edge of the pipe as well as the solder to lie upon each other flush with the general surface, as in Fig. 149. When the plumber lays down the lead he hammers it close upon the hole, puts the piece of pipe through from the top, dresses or beats, that is, tafts the edge of the pipe nicely over the edge of the sheet lead, prepares as usual for wiping, drives in a couple or so of clout nails to keep all in place, and finishes off the tafting with wiped soldering perfectly flush with the level surface of the lead.

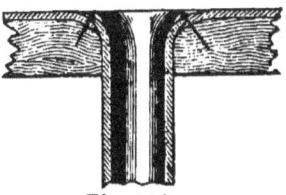

Fig. 149.

Tight Joint.—A sound and reliable connection impenetrable by rain, &c., in roof coverings, but in piping impervious alike to air or water.

Underhand Joint is a wiped soldered joint made on a horizontal pipe brought out 2 in. or 3 in. from the wall for the hand to get round. The mode of making it is fully described under Wiped Joint. It is usual when these and similar pipe joints are made to pipes already in place, to tin where possible the new lengths on the bench previous to connecting them.

Union Joint.—This will be found described in Section XIV. It should be added that elbow and straight ferrule unions are screwed for iron at one end and inserted into water mains as noticed under Ferrule Joint, the other end being tinned for lead service.

Upright Joint is one made on a vertical pipe either by means of the blowing lamp and strip solder, or else with the copper-bit or grosing iron and molten solder. In the latter case a stout paper funnel may be tied round the joint close under it, or, what is much better, a lead collar made out of a circular disc of sheet lead, with a central hole and small sector abstracted for the pipe and its insertion, should be pressed against and clipped round the pipe about 2 in. below the joint. Either of these plans will prevent the solder running off. It is very necessary that the ends of the pipes should be well fitted and secured together before soldering, so that no solder may find its way through to the bore. It is of course also necessary that the proper shaving and soiling be performed. A splashing or pouring stick consisting of a piece of wood with a groove down its centre is often used to splash on and keep the solder well up to the top of the joint, for by pouring from the ladle into the stick, which is held in the left hand, the solder can be directed above the joint in order to fall and flow upon, and raise and maintain, the temperature of that portion of the pipe, which is so essential to thorough tinning, whilst the solder about the collar performs the same good offices lower down. When enough solder has collected and stops round the joint, the application of the iron will melt it again and reduce it to the necessary pliability whilst the wiping cloth is manipulated so as to dexterously fashion it into shape. When a pipe bursts it is not always necessary to cut the pipe and insert a new piece, for by bringing the pipe up again to its original contour with a few taps, and then shaving it and so on as for a new joint, the region of the injury can be brought to the proper temperature for tinning with the splashing stick until enough solder adheres to work and float with the iron, which keeps it flowing whilst the cloth catches and replaces the falling spatterings and finally surrounds the damaged part with a neatly shaped patch.

Wedged Joint occurs when lead wedges, otherwise lead bats, are used to fix sheet lead to grooves, mortar joints, &c.

Welded Joint.—This has been already described under its alternative title of Overlapping Joint. It possesses the advantage over the roll joint of not taking quite so much lead—the height of the two turned-up edges not being necessarily more than $1\frac{3}{4}$ in. and $1\frac{1}{4}$ in. respectively —besides which it forms a less inconvenient and unsightly joint for terraces and the floors of balconies and verandahs, inasmuch as it projects less than the roll above the general level, and is consequently less likely to occasion accidental trippings.

Wiped Joint is so called from being made with the aid of a soldering cloth or wiping cloth formed out of a piece of old or new canvas, fustian, moleskin, bed-ticking, &c., folded into several, say 8 or 9, thicknesses so as to constitute a pad about 5 in. by 4 in., or any other size considered more convenient by the artificer. This is well saturated with tallow by holding it before the fire and rubbing the grease in, or else by placing it in a vessel exposed to gentle heat with the tallow on the top of it, but it matters not what the method is so long as it results in the cloth becoming perfectly soaked, soft, and pliable. The grease, which must be renewed as required, prevents the solder from sticking to the cloth, and imparts the suppleness essential to its proper manipulation by the plumber in order that he may wipe the solder to any form or do, in fact, anything he chooses with it when brought by the iron to the proper buttery consistency. The solder, or pot metal, or metal as it is commonly called in the trade, that is used with this cloth is a coarse variety of soft solder made by melting together 2 parts of lead and 1 of tin, or thereabouts. It must not be allowed to get red-hot, which tends to harden the metal and render it liable to easy fracture, nor must any zinc whatever be permitted to get into the pot, else the whole will become too brittle to work and be spoilt. In order that the solder should not stick where none is required, a mixture of thin glue and lamp black, or some equivalent compound called "soil," "smudge," or "tarnish," is applied with a small brush such as a sash tool for a depth of about 3 in. or 4 in. from the edge of each piece to be joined; but previous to laying it on any grease, accidental dirt, &c., is removed with a piece of chalk, and a rubber of cardboard or thick paper. After the smudge has quite dried by holding near it a hot iron, or by other or natural means, the parts of the work to be united must be prepared and fitted—if of pipes, as explained under Taft Joint—and then for a distance back of about 1 in. or 1½ in. from the intended line of junction they must be carefully scraped with the shavehook clean and bright, and immediately greased with the "touch" or piece of tallow candle to prevent retarnishing. If sheets are to be joined, their edges lying in the chase cut in the boarding for the flush soldering must be similarly prepared, and so must each and every edge that is intended to be united by this kind of joint. The soldering iron being brought to a bright red heat, the solder melted so as to pour off in a bright and silvery looking stream, the scale filed off the iron, the wiping cloth warmed and greased, the ends or edges of the work properly adjusted and firmly held in place, and moreover assuming the joint about to be made a flush soldered one and the weather fine or dry, a ladle full of the metal, not too hot to melt the lead, is poured along over about a foot length of the joint, more metal being really required to raise the work to the proper temperature so that it may become thoroughly well tinned than for absolutely making the joint, the surplus being always, however, returned to the pot. The plumber with the wiping cloth in his left hand presses the solder into place, whilst with the right he works the hot iron backwards and forwards, remelting the solder and bringing it to a soft plastic consistency admitting of being wiped or fashioned at will to any form, and which in this instance is simply a smooth flush surface. In the case of an underhand pipe joint, the wiping cloth is held under the joint and the solder poured in a fine stream along and about the region of the joint, and not all in one place, whilst with the cloth it is pressed around it, notwithstanding much drops and falls off and is skilfully caught by the cloth, the pouring being continued till a sausage-like lump remains. The hot iron is energetically plied backwards and forwards round and about until the solder is brought to the soft state just spoken of, when it is wiped or finished off in the form of a symmetrical knob somewhat egg-shaped, with a thickness of solder at the centre of about $\frac{3}{8}$ in., dying away into the pipe in a well-defined line, as shown in Fig. 150. Sometimes the position of the pipe is such as to require the copper-bit to complete the joint. In making a wiped joint with the blowpipe

Fig. 150.

or blowing lamp and strip solder of the same coarse description as that used as above in the molten state, a single thickness of cloth is all that is required, the flame melting the solder, which drops on the joint and raises the work to the proper heat and the solder to the proper temper, so that the former may be well tinned, and the latter distributed about the joint to enable it to acquire the desired shape and soundness. When enough solder has been deposited on the joint, however, it is still necessary to maintain the heat with the flame whilst the proper symmetrical shape is imparted to the joint with the cloth.

SECTION XI.

ZINCWORKERS' JOINTS.

Capped Joint is made between two sheets and in the direction of the current by means of a roll cap when zinc is used for roofs over boarding. One of the best descriptions of cap, and that adopted by Messrs. F. Braby and Co., the celebrated manufacturers of zinc roofing, consists of one formed out of a strip of sheet zinc and bent down over a wood roll, being confined with clips and fork connections (as further explained in the succeeding paragraph) so as to cover the upstanding edges of the sheets on both sides of the roll, and afford play for expansion and contraction whilst preserving its own position. Drawn roll caps consist of wood rolls with the zinc drawn over them by machinery. These only require to be laid upon the upstanding edges and fixed with screws down upon the boarding, the edges of the sheets fitting into grooves run on the underside of the rolls. In some cases three-quarter round strips stiffened with turned-up edges are slid on without wood rolls, and securely clasp the upstands through their elasticity, but these and other kinds require soldered ends, whereas Braby's zinc and square roll caps with saddle caps and stop ends, which are solid and not soldered, as described in their unique pamphlet on zinc roofing, appear to answer exceedingly well. The joints likewise between the turned-up ends of the sheets and the ridge zinc covering are capped. Other remarks relating to this joint will be found under Zinc Roll Cap Joint.

Clip Joint is one secured with zinc clips, and both the sides and ends of zinc sheets are thus jointed when the plain roll cap is used. The clips are strips about 2 in. wide laid under the wood rolls about 3 feet apart, with their ends turned down to clip the upstands of the sheets. The loose zinc roll cap is then slipped over and along, with fork connections inside soldered on to slip into the hooks of the clips. Such is a plan adopted by Messrs Braby and Co. Clips of similar width are likewise used with the fold joint.

Drip Joint.—This is made in a similar manner to the lead drip, but soldering has to be resorted to at the angles.

Flashed Joint.—This is likewise the same in principle as the corresponding lead joint, excepting that solder is occasionally necessary, and that the edge of the flashing is all the better for being stiffened with a bead.

Fold Joint.—This is shown in Fig. 151, as made by Messrs. Braby and Co. It is used for uniting the upper and lower edges of sheets of zinc laid on boarding, having a slope at least of 1 in 7. Clips of the same metal about 2 in. wide are nailed to the boarding and doubled in between the edges of the sheets, whereby they are secured without too much restraint.

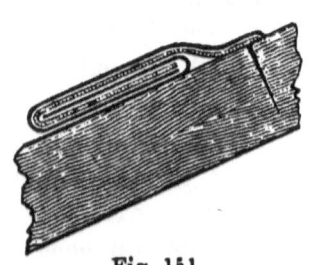

Fig. 151.

Folded Joint is a fixed joint, and not adapted for outside work like the foregoing. It is made by means of a fold, which is hammered and soldered.

Folded Angle Joint.—This is shown in Fig. 152. There are, however, other varieties.

Fig. 152.

Heading Joint is the same as that of the plumber.

ZINCWORKERS' JOINTS. 215

Lap or Lapped Joint.—Besides the description of the similarly termed joint given in Section X., which is more or less applicable to this trade also, it is proper to note that there are other distinct forms, two of which are represented in Figs. 153 and 154.

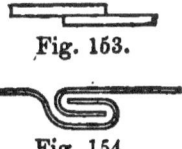

Fig. 153.

Fig. 154.

Mitre Joint occurs when rolls or mouldings are returned, and is made with the copper-bit, as described under that head in Section X.

Roll Joint.—This is described under its alternative names of Capped Joint and Zinc Roll Cap Joint.

S-Joint is used to unite two edges by means of an intermediate strip of S-shaped section soldered to both, as shown in Fig. 155.

Saddle Joint.—This name is given to a joint formed by turning up the edges of two sheets or strips against each other, but one more than the other, and turning the greatest upstand down over the least, as in Fig. 156. It also occurs between a saddle piece at the ridge end of a roll and the ridge capping Formerly it was necessary to solder saddle pieces to the roll caps, but Messrs. Braby and Co.'s patent solid saddle piece dispenses with soldering.

Fig. 155.

Fig. 156.

Screw Joint occurs amongst other positions when drawn roll caps, or zinc sheets on the Italian system of corrugation, are respectively screwed down to boarding or rolls.

Soldered Joint.—The parts where the solder is to fasten on are brushed over with hydrochloric acid (otherwise muriatic acid, or spirit of salt), killed, *i.e.* turned into chloride of zinc with a few small pieces of zinc, and are then soldered with the copper-bit and fine solder, as described under Copper-bit Joint in Section X.

Wedged Joint occurs when strips or pieces of zinc are secured in position at their junction with masonry, &c., by means of wedges.

Welted Joint.—This is synonymous with fold joint, which has been already explained.

Zinc Roll Cap Joint is formed between two sheets of zinc, as already noticed under Capped Joint, of which it is, in fact, another name. A common way of making it is as follows:—A wood fillet, about ¾ in. wide and ½ in. thick, is nailed upon the boarding along the line of junction, and upon this fillet is nailed or screwed a wood roll, about 2¼ in. high and 2 in. wide, with its upper surface rounded and lower part left square, and bevelled off to the width of the fillet. The zinc sheets are laid against this on opposite sides with upstands of 4 in. and 3 in. respectively, which are closely fitted to and bent over the roll, the longest overlapping the shortest; and then a zinc roll cap of cylindrical form is slid on, gripping both sheets where the roll is narrowest and making a snug joint, with ample play for expansion and contraction. This plan requires a piece to be soldered to the roll cap under the saddle at the central ridge, or under the flashing if only a lean-to covering, and also necessitates a short length of roll cap to be mitred and soldered vertically at the edges, or eaves; but Messrs. Braby and Co. cover flats, &c., with rolls provided with solid stop ends, so that their plan almost obviates throughout the use of solder.

SECTION XII.

COPPERSMITHS' JOINTS.

Brazed Joint is the union of the cleaned, brightened, and truly fitted edges or surfaces of brass, copper, iron, and gunmetal, by hard soldering, which is effected by temporarily steadying or binding them well together, and melting between them an alloy called spelter solder, consisting of equal parts of zinc and copper when surfaces of brass are to be joined, and of 3 parts of copper to 2 of zinc when copper or iron is brazed. The former variety is called hard spelter and the latter soft spelter, but they are both technically hard solders. Silver solder consists of 3 or 4 parts of silver to 1 of copper, or 2 of silver to 1 of brass; and it renders some joints—for instance, those of bandsaws—less brittle. These solders are likewise known as hard, but that containing the most silver is relatively the hardest. The spelter solders are used in a coarsely powdered or granular form, the granulation being effected when making the solder by emptying it, after it has been exposed for a few minutes to a glowing heat, into a pail of cold water, and agitating it violently with a broom. They are worked into a mass with borax and rain or pure water, and previous to adding the borax it is advisable to burn it, with the object of driving off its combined water. When thus prepared, the paste is carefully spread or plastered over the surfaces to be united, and the heat applied from the clear fire of a forge with the aid of tongs, or with the flame of a gas blowpipe,—in either case gently at first, so as to evaporate the moisture and prevent the borax and spelter from drying away from the joint. Silver solder is divided into very small square plates, and the powdered borax may be used either dry or wetted with water. On becoming cool, the joint is cleaned off with the file and emery.

Capped Joint.—Where copper is used as a roof covering this is identical with the corresponding joint of the Zincworker.

Copper Roll Cap Joint is an alternative name for the preceding joint, and is made similarly to that of the Zincworker, with solid ends to the roll caps.

Cramp or Cramped Joint.—This is a term employed to designate a mode of joining thin sheets of brass, copper, or other hard metal, by tapering off the edges very thin, scoring a row of notches with oppositely inclined snips along one of them, as in Fig. 157, bending them alternately up and down, as in Fig. 158, and inserting within them the other edge, as shown in Fig. 159. They are then hammered and brazed with borax and spelter, and hammered and filed smooth.

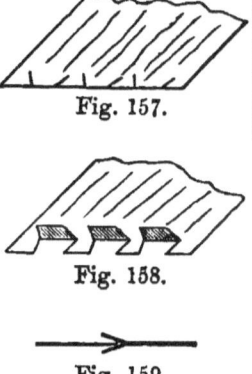

Fig. 157.

Fig. 158.

Fig. 159.

Fold Joint.—This is a similar joint, and made in the same way as the corresponding one described under the Zincworker.

Folded Joint is the same as the corresponding joint in Section XI.

Lap or Lapped Joint.—This is represented in Fig. 160, which shows a mode of forming the joint with thinned edges.

Fig. 160.

Lightning Conductor Joint is used in connecting the

various sections of the conductor, and, as may be imagined, many different forms of the joint have been devised to insure continuity. It can only be observed here, therefore, that as a rule iron should be brazed to other metals, in order that rust may not interfere as an insulator, whilst, perhaps, copper sections are best united either by riveting followed by brazing, or else by giving the ends a cylindrical or circular form, upon which a screw thread is cut, and joining the screw ends by means of a suitable and tight-fitting screw socket piece. The attachment to the building protected is by hooks or staples, and should be as close and complete as possible, insulation being perilous, and to be most carefully avoided.

Roll Joint.—The same as Capped Joint.

SECTION XIII.

GLAZIERS' JOINTS.

Beaded Joint occurs when squares are stopped with fixed or shifting beads attached to the sash with nails or screws. Occasionally they are brass bronzed and moulded, and fixed with brass screws through countersunk holes. In all cases the edges of the glass should be run with black paint previous to stopping.

Bed Joint.—This will be found under various heads. In ordinary work a square of glass is bedded upon the back putty, which is the term applied to the putty first laid in on the rebate. The front putty completes the stopping, and is run along the edge of the rebate on the other side of the glass.

Butt Joint.—This is formed between two pieces of glass which fill a square, but do not lap, and is sometimes adopted in shop windows to save the expense of a large pane.

Cement Joint.—Lights are fixed in grooves in stone by means of oil cement, consisting of Bath-stone dust and boiled linseed oil, the grooves being first sized. Common putty softened with olive oil is sometimes preferred. Portland cement will cause the glass to crack. The cement used for lead lights is a mixture of whiting and white lead, which is dusted over the lead grooves after they have been run over with a thin paint. In the case of iron sash bars a mixture of 1 part tallow and 2 parts resin is laid on after the bars are made hot. Whatever cement, however, is used, whether common putty or any other improved variety, it must not be of a kind to get brittle, owing to the expansion and contraction of the iron sash not being the same as that of glass; and it is also essential that there be no contact between the iron and the glass.

Filleted Joint occurs when panes are secured with fillets or stops, either plain or ornamental, screwed to the stiles or rails, the edges of the panes being painted black.

India-rubber Joint is made by glueing to the rebate a strip of vulcanised india-rubber, and another one to the inside of the bead or moulding, which is screwed on to the frame to secure the square.

Lap or Lapped Joint.—This is common in greenhouses, and is seen sometimes in shop windows where there are no cross bars and the panes are made up of two pieces. It is essential in roofing. Glass tiles and panes of glass in skylights and sloping surfaces always have lap joints at horizontal junctions, bars being inadmissible except in patent systems of glazing without putty, and when the pane is large its tail is connected with the head of that below with a zinc or copper tingle, having oppositely bent ends like those of lead used in replacing slates. When the lap is vertical not much overlap is necessary, but when sloping an inch or more is indispensable to prevent leakage through capillarity or driving winds, and a pointed or rounded tail is preferable to a square one, as it concentrates and expedites the flow of water off the joint.

Lead or Leaded Joint.—This occurs when small pieces of glass are joined together by means of grooves in leaden slips or strips called cames, having an H section so as to form panels or lead lights. Usually a panel consists of a collection of small diamond-shaped panes, held in place by two parallel rows of cames, which cross one another diagonally, and are soldered at their intersections, the whole being surrounded on the outside edge by what is called a broad lead, that is a broader strip than the others, or about ¾ in. in the leaf. The panes are inserted by bending down the sides of the cames and then turning them back again to grip the glass; and in order to keep out the weather the lights should be cemented, as described under Cement Joint.

Open Joint.—Under this head may be named the device for enabling condensed water to run off from the interior of stone windows in churches, chapels, and public rooms. A piece of lead is dressed down over the sill, as shown in Fig. 161, under each light, in a sinking made for it, and the glass is bedded in the grooves so as to leave an opening of about ⅛ in. between its bottom edge and the surface of the lead.

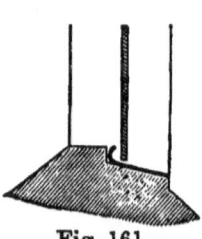

Fig. 161.

Overlapping Joint.—This is another term for lap joint.

Painted Joint.—All joints made with lead putty or oil cement should be painted, to prevent the oil evaporating, and the edges of sheet-glass require to be painted black before bedding. All wooden sash bars, likewise, must be primed or painted before the squares are stopped in.

Putty Joint.—After the woodwork is primed the glass is bedded in putty, which is made of four degrees of hardness, but that generally used is a mixture of whiting, white lead, and boiled linseed oil. There is no universal rule as to the composition of the so-called soft, hard, very hard, and hardest putty. The addition of flour or olive oil makes it softer, and so does the omission of white lead and the substitution of raw for boiled oil. White or red lead, sand, and turps add to its hardness: 7 per cent. of tallow conduces to its continued pliability. In glazing, the back putty is first laid on the rebate to bed the square upon, but this is sometimes omitted both with the best and worst descriptions of glass, but never in work of fair quality. After well pressing the square upon the back putty, the stopping is completed with the front putty, which is run along the edge of the rebate and finished off with a bevelled surface in a direction away from the glass. The front putty is usually on the outside and exposed to the weather, excepting in the case of shop windows, which are glazed from the inside for security.

Puttyless Joint.—This appellation is especially applicable to those joints now employed on a large scale on extensive roofs, which are formed by slipping the glass into grooves to be retained in place by clips, with or without the pressure of india-rubber attached to a capping on the bars. A contrivance in the shape of small gutters in the vertical bars or perforations in the horizontal ones is provided to allow condensed vapour to run off from the inside. Rendle's "acme" system of glazing without putty can be used with iron bars, thereby dispensing with wood, and presenting peculiar and advantageous features for flat exposed roofs. Fletcher's patent "metal substitute" also gives a secure fixing to glass as well as a neat appearance to roofing without the use of putty.

S-Joint is formed in skylights when the panes overlap and the tail of one pane is connected with the head of another by means of an S-shaped clip.

Sprigged Joint.—When the panes are large and heavy the joints between them and the bars are often first sprigged with iron or copper brads or sprigs, which have no heads, and afterwards front-puttied with common or red-lead putty.

Thermo-plastic Putty Joint is formed by adding tallow to glaziers' putty, to enable it to become plastic by the heat of the sun and firm again upon cooling, which quality renders it suitable for securing large panes of glass without the risk of their becoming loose through expansion and contraction.

Wash-leather Joint is made in glazing when large squares in sash-doors, &c., are stopped with plate-glass, in order to lessen the injurious effects of slamming or other sudden concussion. A strip of wash-leather is glued to the inside of the rebate, and another to the inside edge of the bead screwed to the frame or bar to hold the pane in its place. Previous to insertion the edges of the glass must be painted black.

SECTION XIV.

GASFITTERS' JOINTS.

Ball and Socket Joint.—A joint formed by a ball enclosed within and accurately fitting a socket, consisting of about three-fourths of a hollow sphere, the absent portion of which affords space for the tube or rod, attached to the ball, to move freely about in any lateral direction. The socket is made in two pieces, which are either screwed together or else united by flanges and bolts, the plane of junction of the flanges passing through the centre of the ball, so that it can be easily removed when requiring grease. The ball is hollow, and bereft of a portion of its perimeter, as shown in Fig. 162, exactly opposite the corresponding hole in the socket, and in order to obtain free passage for the gas there is a second hole both in the ball and socket opposite those already described, to which the pipes for the exit and entrance of the fluid are respectively attached. Before using for connecting gas pendants the ball and socket must be examined for strength, the ball removed and greased, and the fitting rejected if not air-tight. This elegant form of movable joint is likewise called Ball Joint, and Cup and Ball Joint.

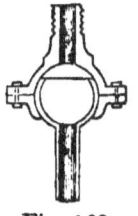

Fig. 162.

Barrel Union Joint is one formed between the ends of tubing or barrel pipe by means of a brass connection called a barrel union consisting of three pieces (Fig. 163), two of which are respectively soldered or screwed to the ends of lead or iron pipes. One of these pieces (A), called the lining, is furnished with a flange near its edge, and the other (B) is provided with a socket, having an outside thread, upon which the third piece (C), called the cap, screws. The cap is a sort of hollow nut of octagonal form, which is slipped over the lining previous to attaching it to the pipe, and when screwed up it brings the flange close against the face of the socket, and by means of a leather washer between them (as indicated by the black lines in the figure) a superior air-tight joint is formed, admitting of being easily loosened when desired. In gasfitting all unions require washers, but in other cases the parts screwed up close need only to be truly ground down to a smooth face.

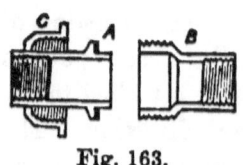

Fig. 163.

Blown Joint.—This is another name for a joint made either with the blowpipe or blowing-lamp, as described in the following paragraph.

Blowpipe Joint is made with an instrument called a blowpipe, which is a tube terminated with a finely pointed nozzle, through which air is forced across a flame to concentrate it upon a particular spot for fusing, &c. The mouth blowpipe usually has a horn, ivory, or wood mouthpiece, and the art of using it consists in blowing and breathing at the same time, or, in other words, in employing only the muscles of the cheeks so as to maintain an even and uninterrupted blast for a moderately long time. Breathing is maintained through the nostrils whilst the mouth is replenished, when necessary, with air without interfering with the blast. Blowpipe joints are chiefly used by gasfitters for joining ¼ in., ⅜ in., and ½ in. pipes, or other small jobs. The work being prepared by carefully fitting the parts to be united, as explained under Blown

Joint in Section X., they are scraped clean, placed in position, well steadied together, and well sprinkled or dusted with powdered black resin. The workman then puts the blowpipe in his mouth, takes a candle with its wick spread open, or what is not so safe or handy, a bunch of burning rushes covered with tallow, in his left hand, and a strip of fine solder in his right. Holding the flame an inch or so off the work so as not to smoke it, he blows a steady, clear flame on the joint, touching it until sufficiently floated, and moving about or along it with the solder, which must flow evenly, and be worked and sweated together without melting the lead, but at the same time so as to enter into close union with its surface, and without letting the solder come off in bits. The æolipile or blowing lamp (Fig. 164) is now preferred to the blowpipe, but its flame is not so powerful. It consists of two spirit lamps, one fixed above the other, both being fed with methylated spirits. When the lower one is lighted it almost immediately boils the other, causing the vapour, which is led round to the flame of the lower lamp by a small curved pipe issuing from its top, to ignite and dart out in a horizontal jet of great heat. The joint made either with this instrument or with the blowpipe may be left nearly smooth.

Fig. 164.

Copper-bit Joint.—As a rule this is not recommended for uniting gas-pipes, but when it is used the same procedure must be followed as described under the Plumber, the essential point being to properly tin the bit and the parts to be joined with a few drops of solder, so as to get perfect union or physical adhesion between the tinning and the work, and after that perfect cohesion between the tinning and the solder which is melted off from the stick or cake with the bit, and dexterously kept hot, distributed, sweated and smoothed by its application.

Cup and Ball Joint.—The same as ball and socket joint.

Elbow Joint is formed with an elbow piece, which may be of brass or iron, and round or square. The latter variety is represented in Fig. 81. Sharp elbows are inferior to easy bends for the passage of gas in branch pipes, &c. Hence lead pipes are in this respect superior to iron, whilst rust often accumulates in the latter and causes inconvenience. A T-piece with one of its branches stopped with a screw plug is a handy substitute for an elbow, inasmuch as it enables water or rust to be removed.

Hydraulic Joint is the same as a water joint. It consists in making a sliding pipe remain air-tight by means of a water lock. Fig. 165 shows the joint as ordinarily used in gasaliers, the outside and inside pipes being parts of the gasalier that are supported by weights or counterpoises and chains suspended from the ceiling, and therefore capable of sliding, whilst the intermediate pipe, which receives the gas from the branch pipe and dips into the water and forms the seal, does not slide but is permanently fixed to the top or ceiling plate of the gasalier.

Fig. 165.

Knee Joint is another name for elbow joint, and sometimes applied to a joint formed by means of an elbow nosepiece.

Pipe Joint.—Iron barrel or iron tubing is united by this artificer in the same way as by the Smith, excepting that washers must in all cases be used between flanges. Lead pipes and compo. tubing are connected up as described under Blowpipe Joint. Cast iron mains are laid by mainlayers similarly to water mains.

Saddle Joint.—In connecting the service with the

meter this occurs as a sort of T-joint, and is made by stopping the end of the lead pipe from the meter and inserting the cap and lining into its side, instead of connecting it end on to the service. In bending the pipe on to meter care should be taken not to contract its fullway.

Screw Joint occurs in uniting wrought iron barrel and fittings to one another, and brass tube to bracket backs, swivels, &c., and all kinds of fittings to straight and elbow nose pieces, &c., &c. The remarks under Socket Joint are here applicable.

Sliding Joint is a sort of telescopic joint, and may be either hydraulic or else formed by means of a cork or hempen stuffing attached to the inner or sliding pipe, to allow it to work backwards or forwards in the outer one, which it keeps at the same time gas-tight.

Socket Joint.—A socket is a short tubular piece either plain, or diminishing, or reducing, with an internal thread which forms a double socket joint between two lengths of tubing when screwed on to their ends with a touch of white lead. A better joint, however, and one that is quite impervious, is made by mixing boiled linseed oil and red lead to the consistency of paint, taking off the socket, and painting over the screw ends of the tubing or barrel, then binding a thread of raw hemp round the screw, and finally screwing up all tight.

Soldered Joint.—This will be found described under Blowpipe Joint, Copper-bit Joint, &c.

Stiff Joint is, as its name implies, a fixed joint, and in gasfitters' work signifies one that does not revolve like the ball or hang loosely like the swing joint.

Swing Joint occurs either in a swing pendant or in the ordinary single, double, or treble joint swing bracket. It is another name for the swivel joint described in the following paragraph, and is often used in piping when lateral motion is required.

Swivel Joint.—This consists of a perforated plug moving in a sort of shell, and capable of turning completely round on it in one plane. It is the joint which affords motion in ordinary gas brackets. The plug is only kept in its seat by a screw, upon which too much pressure should not be permitted to fall. There are single, double, and triple swivel joints in common use, and it is somewhat singular that stops to prevent the burner from approaching too close to the wall are not more frequently attached to them. Swivel joints are liable to become more or less choked with tallow or white lead whereby the pressure of the gas at the burner is interfered with, and if the poverty of the light cannot otherwise be accounted for the joints should be taken to pieces and cleaned.

T-joint.—This has been described under the Smith, when it occurs in a main. In gas-fittings, whenever a new connection has to be made with a service, there is, perhaps, less chance of crippling it by inserting a proper T union in the way described under Union Joint, than by joining on the new pipe by merely soldering the two together, which is effected in much the same manner as explained under Branch Joint in Section X. The best way to insert a service into a main is to use the drill and drill clutch; then rimer, if necessary, and tap.

Tight Joint.—This, of course, is the result of good materials and good workmanship at the joint, added to careful bedding and fixing to prevent strain from disturbing the work. In the case of mains, leaky joints probably spring far more from settlement and vibration from traffic than from expansion, &c., from changing temperature. Gas pipes should not be let into walls, for, besides any corroding action that might set in, the joints are liable to injury from strains brought upon them through expansion or settlement. Small wrought-iron barrel for branches is not always approved of, because much rust sometimes accumulates in their interior and causes annoyance. Copper pipes are subject to the formation of a detonating compound within them, through the action of gas on the metal. Great care is needed in attaching heavy gasaliers, though with clean, accurately cut threads there is little chance of shearing occurring at the screw joints, the most likely accident arising from the bridging or bridgewood starting, or framing settling, and throwing a cross strain upon the stiff joints. Sometimes in screwing on a small pendant to a brass nose-bit the operation loosens the soldering at the other end, and if a short length of pipe, screwed at one end for the pendant and tinned at the other for soldering to the branch, be used instead, the same danger exists unless great care is shown in screwing up, for though there may be no leak at first, the traffic overhead may soon extend the flaw. The joints and fittings are tested by means of a small disc attached to the meter, the index of which should remain stationary, with the gas turned on at the meter but off at all the burners. The connections may likewise be tested by sucking the main or attaching an air-pump. Leakage sometimes arises from the arms of gasaliers being made with too thin metal which is still further reduced round and about the joints. To guard against this and similar dangers it might be advantageous to test fittings under a pressure equivalent to that of gas when balancing a column of water 2 in. in height, which is about double that found in gaspipes between sunset and midnight in many towns. A mixture of 7 parts of air and 1 part of gas is a fearfully explosive compound when ignited; hence the slightest odour of gas should be traced to its source and the leakage stopped, but if the odour be strong its neighbourhood ought not to be approached by a light unless there is every reason to believe that the proportion of air to gas is more than 11 to 1. Though the diffusion of gas is rapid, it would be sometimes dangerous to apply a light to the ceiling of a room when it could be held lower down with impunity.

Union Joint is a mode of uniting the ends of pipes by means of a coupling, which always consists of not less than three pieces, two of which are usually either screwed or soldered to the ends of the piping about to be united, whilst the third piece, which is a 4, 6, or 8-sided cap or nut, screws them both up tightly together, a leather washer being always interposed between them. A variety has already been described under Barrel Union, which embodies the principle of all those most in vogue for pipes. Brass and iron unions are either screwed for iron pipes or tinned for lead at both ends, or one end is treated each way. Union T's, nipple and ferrule unions, are similarly prepared, the former having a cap and lining to two out of its three branches or arms, the third being directly screwed or soldered to one of the pipes, whilst the two linings are attached to the remaining two pipes, the whole three-way junction being connected up and completed by means of the caps. A nipple union is sometimes screwed into an iron service instead of a T.

Universal Joint.—The cup and ball and the universal swivel are varieties of this joint, for both of which, however, flexible tubing is sometimes found a convenient substitute when of impervious texture, though it is hardly ever reliable for the transmission of gas for any length of time.

Water Joint.—This is another name for the hydraulic joint.

SECTION XV.

PAPERHANGERS' JOINTS.

Butt Joint is usual with lining paper the edges of which do not lap but meet at a butt. The best papers are also edged, so that both margins are cut off close to the pattern in accurately straight lines, the pattern not admitting of any but neat butt joints, which should be sufficiently perfect to be invisible.

Close Joint is the customary joint resulting from one or both of the edges being cut close to the pattern and the paper neatly hung. It is produced either by an exact butt or lap.

Lap Joint.—This is formed by the overlap usually given to common and medium papers which require one edge, and generally the left-hand one, to be cut close, whilst a margin of ¾ in. or thereabouts is left on the other for the former to cover. In hanging the paper the overlap should be so given that, where possible, the cut or upper edge may be turned away from the light.

Mitre Joint.—This occurs in panelling, bordering round doors, &c., the angles of which should show a clean-cut mitre.

Pencil Joint.—This is made by lining and jointing marble paper by running lines of distemper colour to represent blocks of the desired dimensions. The joints may be made to look as if belonging to ashlar, or arch work with central key.

Rolled Joint is effected by rolling the overlap with a wooden roller covered with flannel.

SECTION XVI.

PAVIORS' JOINTS.

Asphalte Joint.—This is used in street paving, the granite sets being laid and bedded on a coating of about 3 in. of gravel, spread over a foundation of hard material about 1 ft. in thickness. The joints between the sets are then filled in with asphalte, or a composition of creosote, pitch, and tar, in proportions varying according to the quality of the pitch, one variety of the mixture consisting of these constituents in the respective proportions of 1, 11, and 4 by weight. After boiling for a couple of hours the mixture is poured into the joints, consolidating the pavement and preventing the deposition of mud under the sets. With the same mixture blocks of wood or stone may be bedded and jointed to form paving, and joints in roadway planking under granite in iron bridges rendered water-tight, after all fissures and cavities have been stopped by caulking.

Bed or Bedding Joint.—Previous to bedding the ground must be made up, consolidated, levelled, and covered with gravel to a depth of 3 in., more or less. Flags are bedded in mortar, paving bricks in mortar, and then grouted over and between the joints. Common tiles are laid in mortar upon a bed of lime core, brick rubbish, or coal ashes, not less than 3 in. thick, and grouted with cement. Superior tiles are laid upon a bed carefully floated, and laid to the proper fall, and grouted with cement, the excess being wholly washed off, so that none may be left to set on the surface. York or other stone paving, squared to shape, is bedded on mortar or cement, and set with joints as close as possible to preserve the courses. Kerbstones, gutter soles, and breast stones should be bedded on six inches of concrete, or more if on loose ground or where necessary, and the whole properly bonded, with broken and close joints well grouted and flushed solid, hydraulic lime only being used. Granite sets are bedded in first-class roads with close joints upon an inch or two of sand, overlying a bed of concrete some 12 in. thick. The concrete rests upon brick rubbish about 8 in. deep, rolled and bound with sand as a bottoming. Sets used for crossings require a thin layer of sand and 6 in. of concrete for a bed. Pitching on sea slopes is bedded on a layer of clay and small stones mixed, 12 in. thick. On sand the puddle requires protecting by a bed of rubble 18 in. thick.

Close Joint.—One that is as close as possible consistent with the preservation of the courses and the irregularities of the edges and sizes of the stones, &c. In many instances, where grouting is depended on for filling up all crevices and consolidating beds and joints, it is fortunate that the roughness of the stones precludes too close a joint, which is rarely less than $\frac{1}{4}$ in. when sets are used. Wood paving, to be safe, requires close joints not less than $\frac{3}{8}$ in. in width, otherwise dirt accumulates within them and renders the surface dangerously greasy.

Gravel Joint.—This occurs in paving with granite sets, which, after being laid as closely as possible, are covered with clean, screened, dry, hard gravel, well brushed and worked into the joints. The sets should then be rammed and more gravel similarly applied, and the operation repeated again if necessary to attain thorough compactness and solidity. The same kind of joint does well for carriage-way paving in yards, &c., composed of larger stones, laid close and bedded on a few inches of rough ballast.

Grouted Joint.—The joints between granite sets or cubes are grouted with thin hydraulic lime grouting, or liquid mortar, after ramming, and then covered with a coating, a full inch thick, of fine clean gravel. Wood pavement is sometimes grouted with hydraulic lime and sand, the joints being occasionally partially filled with asphalte previous to grouting, or the joints between wood blocks impregnated with creosote may be run with a mixture of pitch and creosote in the proportion of 1 cwt. of the former to 1 gallon of the latter, alternated two or three times with coats of fine gravel.

Racked Joint is produced by filling up the joints between paving sets with small broken stones or fine gravel.

INDEX.

ABSORPTIVE power of stone, 66, 94
Abutments, bridge, 46, 111, 130, 134; in carpentry, 78, 81, 84, 86, 88
Abutting—cheeks, 95; joint, 78, 100, 105, 106, 111
Action of waves on masonry, 42, 44, 55
Adhesion, 65, 169, 174
Adjustable joint, 78, 111, 141
Adjusting joint, 112, 130
Æolipile, 227
Agricultural drain-pipes, 7
Air—holes, 188; space, 4, 55, 80, 176; tight joint, 7, 9, 12, 200, 225, 226, 228, 229
Anchor and collar joint, 118; anchor bolts, 38; anchor joint, 113
Anchoring plate, 113, 152
Angle—bands, 79; bond, 15; iron, 114, 124, 142; iron joint, 113; joint, 79, 91, 114, 157, 178; of repose, 57, 125; posts, 79; of buildings, 41, 53
Angular—grooved and tongued joint, 79, 157; iron laths, 67; joint, 11, 34, 79, 157, 181
Apertures, 23
Apex stones, 55
Apron or apron flashing, 57, 194, 195, 198
Arch, 56, 57
Arched—beams, 103; flooring, 54; principals, 38; ribs, 111, 112, 114, 130, 134
Architrave, 35, 52, 55, 124, 166
Arch—joint, 11, 34, 51, 55, 114; ring, 53; work, 18, 22, 25, 26, 46, 50, 52, 55, 60, 233
Area, aggregate of headers, 40
Arris, 37, 43, 49, 81, 97, 111, 157, 172, 173
Ashlar, 39, 52, 55, 57, 64, 65; facing, 35, 43, 46, 54, 61; quoins, 54
Asphalte, 9, 10, 111, 145, 155; joint, 6, 35, 234; asphalted felt, 71, 80, 115
Astragal, 184, 204; joint, 184, 193
Autogenous soldered joint, 184
Axed—joint, 35, 47; work, 35, 48, 59, 62
Axis of pillars, 36, 116

BACK, 11, 97; joint, 12, 19, 25, 31, 35, 65, 67, 71; pieces, 100; pointing, 69, 76; putty, 220, 223; rebate, 34, 57; rebate joint, 35
Backing, 28, 40, 46, 49, 68, 153, 171

Bad—forging, 156; gas-light, 230
Bag joint, 36
Balcony floors, 52, 197, 209
Ball and socket, 141; joint, 225
Ballast, 236
Balusters, 54
Banded—architrave, 60; joint, 79, 107
Barff's process, 119, 144
Bar, 126; iron, 81, 142
Barrel, 9; pipe, 147, 225, 228; union, 139; union joint, 225
Basement, 60
Base—plates, 38, 116; stones, 36
Bastard—ashlar, 54; tuck-pointed joint, 12
Bats, 25, 58, 198, 209
Battening, 33
Battens, 68, 71, 75, 77
Battering work, 51, 57
Batting, 59
Bayonet joint, 114
Bead or beading, 160, 214, 220
Bead joint, 12
Beaded joint, 157, 220
Beam, 55, 97, 104, 109; filling, 23
Bearers, 109, 110, 136
Bearing—blocks, 37; plates, 92, 112, 115, 122, 132, 184; rollers, 113, 134; surface, 13, 25, 37, 38, 59, 78, 80, 102, 108, 113, 123, 131, 140
Bed or bedding joint, 6, 12, 18, 22, 36, 41, 51, 56, 59, 67, 71, 80, 115, 131, 185, 220, 234
Bed stones, 36, 115, 122, 136
Bedding, 13, 22, 40, 65; pipes, 201
Bed-dowel-joggle, 47
Bed—plate, 111, 112, 115, 122, 123, 130, 134, 137, 140; plug, 47, 56
Beds, 13; and joints, 43, 44, 59
Beeswax, 42, 155
Bell pulls, 171
Bend, 139, 228
Bending, 133
Beton, 58
Bevel or bevelled joint, 14, 80
Bevelled—halving, 90; shoulder joint, 81; washer, 151
Binder or binding joist, 40, 78, 80, 97
Birdsmouth, 11, 14, 35, 37, 79, 81
Black—cement, 42; mortar, 32; or blue pointed joint, 14; putty, 26
Block, 13, 54, 163, 166; joint, 185; in-course masonry, 39, 44, 65

238 INDEX.

Blowing lamp, 207, 211, 227
Blown joint, 38, 166, 226
Blowpipe, 226 ; joint, 226
Blue—heat, 84; lias lime, 24, 36; mortar, 32, 56
Boarded lining, 87
Boarding, 33, 74, 114
Boaster, 48, 59, 63
Boasting, 36, 63
Bolt, 13 ; heads, 82 ; heads and nuts, 83, 102, 132; holes, 82, 118, 130
Bolted joint, 38, 90, 91, 101, 115, 117, 126, 127, 130, 131
Bolted and checked joint, 109
Bond course, 56
Bond or bonding joint, 14, 23, 25, 39, 52, 65, 68, 70, 72, 74
Bond timber, 13
Booms, 111, 117, 119, 127, 137, 138, 140, 154
Borax, 218
Bordering, 233
Borders to hearths, 168
Bossing up, 191
Bottle nose, 187 ; drip joint, 186
Bottoming, 235
Box girder, 122
Brace, 84, 96 ; joint, 84, 119
Bracing, 119, 124, 126
Bracket, 118 ; back, 229
Branch—joint, 187, 206 ; pipe, 187, 228
Brandering, 183
Brass-fittings, 187, 190, 217
Brazed joint, 119, 217
Breakwaters, 38, 46
Breast stones, 235
Brick, 18, 22, 29, 169 ; arch, 17 ; drain, 6 ; dust, 53 ; fillet, 20 ; nogging, 23 ; piers, 26 ; sewer, 6, 7; work, 13, 114
Bridge and bridgework, 57, 61, 106, 117, 118
Bridge—floors, 90 ; of Bamberg, 93; ribs, 130 ; truss, 138 ; wood, 231
Bridging joists, 97
Bridle joint, 79, 84, 97
Brittle putty, 221
Broached work, 35, 44
Broad—axe, 59, 63 ; lead, 222
Broken joint, 14, 27, 41, 68, 72, 119, 158, 181
Bronze cramps and ties, 45, 60
Brooming, 26
Brown paper, 121
Brushing and wetting, 56
Buckling, 133, 160, 170
Built—beam, 86, 103 ; up girder, 120
Bulging, 113
Burnt—clay, 53; in joint, 187 ; rivets, 143
Burr, 117, 141, 187, 207
Burst pipes, 208
Burt and Pott's casements, 155
Butt and mitre, 79, 107
Butt angle joint, 121

Buttered bricks, 21
Butt joint, 7, 8, 19, 28, 41, 73, 81, 89, 120, 128, 134, 136, 137, 158, 178, 181, 188, 220, 233
Bye pipe, 151

CAISSONS, 85
 Calf, 101
Calked joint, 86, 122
Cames, 221
Caoutchouc, 135
Cap, 13, 115, 126, 131, 132 ; and lining, 226, 229, 232 ; plates, 116 ; sills, 109
Capillary attraction, 192, 221
Capital, 116
Capped joint, 213, 217
Capping, 68, 117, 177
Carbon, 80, 94
Carpenters' boast, 85
Carrying up, 28
Case hardened, 118
Casement, 176, 178
Casing, 37, 174 ; blocks, 49
Casting of joiners' work, 170
Castings, iron, 38, 114, 116, 118, 121, 123 133, 137, 138, 155
Cast iron, 60 ; beams, 118 ; bed plates, 117, 140 ; cramps, 64 ; cylinders, 133 ; dowels, 47 ; joint, 121 ; mains, 139, 147, 228 ; nails, 68 ; ornamentation, 61 ; piles, 150 ; pipes, 9 ; segments, 114, 131, 154 ; sewer pipes, 9, 109 ; sheet piling, 142 ; standards, 119 ; voussoirs, 141
Cast lead, 139 ; sockets, 184
Cauked joint, 86
Caulked joint, 41, 85, 86, 121, 188
Caulking, 6, 109, 115, 122, 125, 137, 151, 155, 234; iron, 85, 122, 135 ; plates, 80
Ceiling, 183 ; joists, 197
Cement, 9, 13, 18, 36, 47, 55, 133 ; bond, 15, 17, 32 ; fillet, 20 ; grout, 22, 37 ; joint, 7, 41, 64, 68, 72, 121, 188, 220 ; mortar, 16, 17
Centerings, 100, 110
Central pivots, 101
Centre, 34, 83, 102, 103 ; pins, 140
Cesspool, 207
Chain—bar, 48, 63, 124 ; bond, 41, 42, 63, 124 ; riveted joint, 122 ; riveting, 122, 127, 128
Chairs, 38, 117, 118, 128
Chalk line, 26, 31, 32
Chamfered joint, 43, 123, 158 ; chamfering, 60, 123
Channelling, 60
Chase, 25, 27, 86, 177 ; mortise, 100 ; mortise joint, 86
Chasing, 58
Checked, 58 ; joint, 43, 86, 101
Check plate, 92, 105
Cheeks, 74, 81, 86, 88, 95
Chimney, 23, 72, 194 ; bar, 11, 122 ; bond, 15 ; can, 17 ; piece, 35, 45, 175 ; pot, 32 ; pot joint, 17 ; walls, 30

INDEX. 239

Chipped and filed, 111, 116, 131; joint, 123
Chips, 48
Chisel drafts, 43
Chiselled work, 59, 62
Chloride of zinc, 190
Circular—framing, 164; gratings, 115; joint, 43, 86, 123; sunk work, 43; work, 57
Cistern, 133; cover, 178; joint, 72, 158, 188
Clamp—or cramp, 159; bricks, 29
Clamped joint, 43, 124, 159
Clay, 7, 8, 9; joint, 7, 185; luting, 197
Clean deal, 93
Cleaning—off, 22; brickwork, 30; old joints, 26
Cleansed joint, 59
Cleat, 97, 114
Clench or key, 182
Clifton Suspension Bridge, 130, 146
Clip, 114, 124, 139, 199, 213, 214, 223; joint, 124, 214
Close—boarding, 71, 87; cut joint, 72; cut and mitred joint, 72; joint, 13, 17, 43, 72, 86, 124, 159, 233, 235
Closers, 15
Coach screws, 85, 105
Coal—ash mortar, 56; dust, 14, 74; plates, 58
Coarse—joint, 17, 44; solder, 205, 206, 210
Cocked—bead, 165; joint, 86
Cofferdam, 83, 85, 87, 93, 102, 103, 120, 153, 154
Cogged joint, 86
Cogs, 67, 85
Cold, 133; chisel, 123
Collar, 7, 8, 17, 113, 117, 124, 125, 126, 139, 152, 201; joint, 8, 11, 18, 124
Coloured—mastic, 65; mortar, 54
Columniation, 41
Columns, 52, 119, 126, 149, 154, 155
Common—halving, 90; joists, 80; papers, 233; rafters, 81, 97
Composition tubing, 197
Compound joint, 2
Compressive stress, 89, 102, 121, 136
Concrete, 6, 22, 31, 37, 38, 41, 49, 115, 132, 133, 135, 153, 155; floor, 152
Conductivity of stone, 94
Cone joint, 188
Conical openings, 57
Connection, 4
Contraction, 133, 144, 184
Coping, 23, 42, 47, 50, 56
Copper, 61, 81, 217; bit, 203, 211, 215; bit joint, 189, 227; clips, 114; cramps or ties, 45, 60; dowels, 47; gas-pipes, 231; roll cap joint, 218; slating nails, 77; tongue, 73
Corbel, 94
Core, 132, 199, 203; of lime, 24
Corked joint, 87, 125
Corking, 13, 46, 54, 63, 85, 109, 122, 151

Corner stones, 41
Cornice, 59, 61, 65
Corrosion of iron, 45, 145
Corrugated iron, 114, 119
Cottered joint, 87, 112, 125, 155
Cotters, 79, 82, 98, 105
Counterforts, 41
Counterlathing, 183
Countersinking, 128
Countersunk—holes, 13, 220; rivets, 144
Coupling, 126, 139, 154, 232; joint, 126, 136; link, 117
Coursed—ashlar, 39; masonry, 49; rubble, 55
Coursing joint, 18, 21, 45
Cover—joint, 127, 148; plates, 120, 122, 128, 134; stones, 80; strips, 127
Cracks, 7, 18, 55
Cramping, 40, 180
Cramp, 35, 42, 60, 61, 145, 151, 159; holes, 202; or cramped joint, 45, 127, 159, 218
Cranked, 131, 135, 142, 154
Creosote, 234, 236
Cresting, 67
Crosette, 47
Cross—bars, 135; bracing, 118, 119, 151; bond, 39, 40; grooved and tongued joint, 159; grooves, 166, 169; heads, 109; joint, 14, 18, 19, 25, 30, 46, 64, 65, 87; section, 120; strain, 103, 107, 115; tongue, 167; walls, 41
Crossing, 16, 30, 173
Crown, 34, 47; tiles, 67, 69
Crushing, 6, 102, 128, 144, 174; strength of mortar, 28
Cup and ball joint, 228
Curb, 104, 105, 109, 115; plates, 100
Curved ribs, 83, 91, 100
Curve of equilibrium, 53
Curvilinear batter, 57
Cushion rafters, 93
Cut—bricks, 15, 27; joint, 20, 21; slates, 73; and gauged arch, 19; and rubbed work, 20; and struck joint, 19
Cutting punch, 127
Cylindrical—castings, 131, 132; riveted work, 127

DADOS, 165, 166, 170, 181
Damp, 93, 104; course, 23, 35
Deflection, 113, 116
Deformed bricks, 17
Diagonal—bars, 127; boarding, 77; bond, 15; bracing, 124; joint, 19, 127; ties, 126, 148, 152
Dipped joint, 19, 20
Discs, iron, 152
Disposition of rivets, 120
Distance pieces, 119
Distemper, 233
Distribution—of pressure, 34; of strain, 78
Dock—flooring, 50; gates, 113, 140; wall, 46
Dog, 118; irons, 85

INDEX.

Dog's—ear joint, 191; tooth bond, 39
Domes, 57, 83, 106, 125, 199
Door—case or frame, 13, 38, 58, 100, 175; head, 162; post, 54
Dormers, 72, 74, 75
Double—cone joint, 191; dovetail cramp, 45; dovetail dowel, 47; dovetail plug holes, 56; herring bone paving, 19; lapped joint, 160; notch, 108; notched joint, 87; nut joint, 127; nuts, 150; quirk bead joint, 160; riveted joint, 127; riveted butt joint, 128; riveted lap joint, 128; shear, 120, 138, 149; springer, 61; tenon, 96, 169; tongued joint, 160; tusk tenon, 80
Doubling eaves course, 72
Dovetail—bolts, 38, 126; joint, 19, 46, 87
Dovetailed—and housed halving, 90; backings, 160; halving, 90; notch, 38; tenon, 96
Dovetailings, 19, 40, 107
Dovetails, 19, 85, 108, 167
Dowels, 13, 42, 45, 46, 56, 59, 60, 61, 97, 130, 153
Dowel-joggle, 52
Dowelled—floor, 177; joint, 28, 46, 68, 128, 100
Downpipes, 117
Draft, 36, 43, 55
Drain-pipes, 7, 185
Draw or drawing, 9, 56, 88, 89, 90, 128, 179
Drawbore pin, 99
Drawboring, 99
Drawers, 160
Drawn—joint, 19, 20, 21, 23, 191; lead pipe, 196, 203; roll cap, 213, 215
Dressing, 13, 35, 37, 48, 55, 59, 62, 63, 117
Drift, 141, 143
Drill, 131; clutch, 230
Drilled joint, 128
Drilling, 128, 140
Drip—joint, 191, 214; step, 186, 191
Driving — ferrule, 132; plugs, 171; punch, 131; wedges, 109; winda, 76, 221
Drought, 5
Drowned lime, 22, 50
Drums, 37, 47, 59, 201
Dry—bricks, 22; timber, 80; weather, 105
D-trap, 204
Dust, 160, 170
Dutch arch, 11

EARS, 67, 118, 124, 139
Easing centerings, 110
Eaves, 71; course, 72; gutters, 122; troughing, 119
Edge-nailing, 173
Effective—area, 102; diameter of screw or screw bolt, 82, 119
Elastic joint, 192

Elbow, 139; joint, 47, 129, 192
Electric insulation, 219
Embossed bed joint, 34
End joint, 47, 62
English bond, 15
Enrichments, plaster, 182
Entablatures, 125
Equality of bearing, 107
Even distribution of pressure, 52
Expansion, 3, 7, 8, 123, 151, 153, 221, 224; joint, 129, 152, 192; rollers, 130, 134
Exposed situations, 98, 129, 155
Extrados, 11
Eye, 106, 124; joint, 130

FACE, 14, 46, 47; joint, 20, 47, 60; nailing, 173; stones, 56; work, 26
Faced—joint, 131; surfaces, 121
Facial plane, 62
Facing, 17, 28, 40, 155; blocks, 49; bricks, 29; joint, 28
Fair—axed joint, 47; joint, 36, 47
False joint, 20, 30, 48, 181
Fangs, 46, 122
Fat lime mortar, 49
Faucet, 8; joint, 131, 193
Feather, 115; edged boarding, 166; or feather tongued joint, 161; piles, 154; wedged joint, 88, 161
Felt, 77
Fence walls, 22
Fender-pieces, 104
Ferrule, 131, 147; joint, 131; union, 232
Fibrous plates, 98
Filing down, 123
Filings, 146
Fillet, 20, 73, 75, 133, 221
Filleted joint, 20, 48, 68, 73, 221
Filletings, 68, 74
Filling, 9, 37, 40
Fillistered joint, 161
Fine—joint, 11, 17, 20, 48; mortar, 37, 44, 48; plaster, 68; sand, 63; solder, 205; tooled joint, 48
Fir—laths, 70; pins, 67
Fireproof floor, 134, 138, 182
First-class roads, 235
Fish—or fished joint, 88, 132; plates, 89, 104, 122, 126
Fished beams, 83
Fishing pieces, 89, 114
Fixed joint, 2
Fixing window frames, 65, 180
Flagging, 61
Flags or flagstones, 13, 234
Flanched joint, 68
Flanches, 118
Flanching, 32, 69
Flange, 13, 75, 111, 131, 132; joint, 132, 193
Flanged—segments, 125; spigot, 116
Flannel roller, 233
Flashed joint, 194, 214
Flashing, 20, 74, 198

INDEX.

Flat, 199; bearing, 123; joint, 20; joint jointed, 20; pitch, 76; ruled joint, 20; struck and jointed joint, 20
Flaw, 143, 156
Flemish bond, 15
Fletcher's metallic substitute, 223
Flexible—gas tubing, 232; joint, 125, 194
Flitch, 102, 105, 119; beam or girder, 80, 83
Float, or flow joint, 194
Floating solder, 186, 190, 191, 195, 203, 227
Floor boards, 158, 159
Flooring, 50, 177; clamp, 159
Flower of sulphur, 135
Flue pipe, 41
Flush—joint, 21, 23, 45; jointed joint, 21, 22; soldered joint, 194, 211
Flushed joint, 21, 29, 48, 52, 60
Flushing, 15, 19, 21, 22, 31, 37, 49, 64
Flux, 186, 190, 205, 206, 217, 227
Fold joint, 214, 216, 218
Folded—angle joint, 133, 214; floors, 158; joint, 214, 218
Folding—joint, 161; keys, 86; wedges, 101, 110, 125
Foot plate, 123
Footing pieces, 38, 78
Footings, 13, 116
Force pump, 139
Forging, 105
Fork connections, 213, 214
Forked ends, 126
Forks, 151
Foundation, 36, 55, 57, 87, 109, 115, 155, 234; plates, 116; slabs, 132, 137; stones, 36
Fox wedged or foxtail joint, 89, 161
Fracture, 6, 144
Frame or framing, 81, 92, 95, 96, 99, 104
Framed, 78; floors, 83; joint, 89, 161; partition, 80, 83, 96, 106
Franked joint, 161
Franking, 161
French arch, 11
Frenchman, 31
Fretwork, 41
Friction, 47, 171, 173, 185; rollers, 113
Frogs, 30
Front putty, 220, 223
Frusta, 37, 201
Full—butts, 121, 152; joint, 21, 27, 28, 49
Fullway, 9, 150
Funnel, 186, 208
Fusing point, 135, 206

G ABLES, 55, 194
Galvanised—bolts, 119; iron, 145; iron cramps, 45; iron roofs, 136, 141; wall ties, 16
Garden wall bond, 15
Gasaliers, 228, 231

Gasket, 6, 121, 200
Gas—blowpipe, 184, 218; bracket, 230; explosions, 231; fittings, 229, 231; pipes, 227, 231; pressure, 231; service, 230, 232; tubing, 145, 151
Gate, 83, 200; posts, 101
Gauge, 23, 30, 70, 71; rod, 31
Gauged—arch, 11, 20; cement, 16, 42; work, 17, 19, 26
Gemmels, 165
Gibs, 125; and cotters, 78, 126
Girder, 13, 55, 78, 94, 116, 117, 122, 130, 153; bed, 117; work, 120
Gland nut, 189, 191
Glass tiles, 67, 221
Glazed tiles, 69
Glaziers' putty, 48, 222
Glazing stone lights, 220
Glued—and blocked joint, 163; or glued up joint, 162; wedges, 180
Gothic arch, 35, 55
Grain, 96, 108
Granite, 35, 43, 47, 49, 58; axe, 48; bedding blocks, 117; sets, 234
Gravel, 236; joint, 235
Grease, 9, 201
Greasy roads, 235
Green—brickwork, 18; copperas, 30; houses, 221; timber, 106
Grit, 54
Groining, 181
Groove or grooving, 23, 58, 60, 66, 88, 89
Grooved—abutment, 84; and feathered joint, 163; and rebated joint, 163; and tongued joint, 28, 50; joint, 49, 134, 181; tongued and beaded joint, 164; tongued and mitred joint, 164
Grosvenor Bridge, 52
Grounds, 157, 171, 181
Grouped riveted joint, 134
Grout or grouting, 21, 22, 31, 41, 59, 65, 235, 236
Grouted joint, 21, 50, 236
Guard—bars, 202; piles, 39
Gunmetal, 217; cramps and dowels, 61, 81; tongue, 50
Gusset plates, 142
Gutter, 32, 205; boards or boarding, 86, 207; bolts, 119; soles, 235
Guttering, 119

H ACKED out joint, 22
Hacking out, 26
Haired lime, 75
Hair mortar, 67, 73, 74, 75
Half—lap joint, 90; mortise and tenon joint, 164; rounds, 75; slating, 74; socket joint, 8; timbered houses, 79
Halved—joint, 88, 90, 93, 103; and bolted joint, 90
Halving, 87, 88, 90, 93, 101; and pinning, 79; in, 142
Hammer, 35, 109, 121; headed key joint, 164
Hammered rivet, 144
Handrailing, 158

M

INDEX.

Handrail—joint, 164; screw, 164
Hand riveting, 143
Hanging, 165; papers, 233; steps, 56; stiles, 140
Hard—solder, 205, 217; soldering, 149
Harris's iron windows, 137
Hatchet bolt, 189
Haunches, 34
Head, 103, 143; of a slate, 74
Headers, 15, 35, 40
Heading—bond, 15; joint, 22, 47, 50, 73, 75, 90, 100, 165, 195, 214
Hearting, 15, 28, 40, 41
Hearths, 54
Heel—post, 113, 140; strap, 106
He joggle, 43, 50
Hemp, 229; packing, 194
Herringbone—bond, 15; paving, 19
Hewn—ridge stones, 72; stone, 44
High—joint, 22; pitch, 70; winds, 130
Hinge, 126; joint, 134, 139, 165
Hinged—joint, 173; shoes, 134, 148
Hinging, 113, 158, 160
Hip, 71, 72, 74, 195; hooks, 67, 68; tiles, 67, 68
Hold, 22, 182
Holding down bolts, 38
Holdfasts, 171
Holes, 88, 129
Hollow—joint, 22, 23, 51; roll, 199; roll joint, 195; walls, 16, 32
Honeycombing, 145
Hook, 134; joint, 16, 134, 165
Hoop, 89, 93; iron, 30, 146; iron bond, 15, 17, 124, 136
Horizontal joint, 25, 51, 60, 135
Hot—oil, 93; water pipes, 122, 139; water pipe joint, 193
Housed, 53; and dovetailed joint, 91; joint, 51, 91, 165
Housing, 51, 166, 175, 180
Hydraulic—concrete, 41; joint, 228; lime, 6, 24; pressure, 139
Hydrochloric acid, 190

I GIRDER, 151
Impact, 3
Imperfection of labour, 10, 29
Inbonds, 40
Inclined—bed, 51; joint, 91
Indented joint, 91
Indenting, 101
Indents, 62, 102
India rubber, 121, 139, 204, 221, 223; joint, 135, 166, 194, 221; packing, 193
Insecurity of cast iron, 130, 133
Inside—joints, 21; lining, 166
Interties, 134, 138
Intrados, 17, 19
Invert, 23; blocks, 6
Iron—band, 124; barrel, 228; bars, 38, 51; bridges, 118; castings, 38; cement, 42, 115, 135; cement joint, 185; chairs, 38; cleats, 143; columns, 13, 36, 54, 115; cramps, 35, 45, 145; cylinders, 37; dowels, 47, 128, 130; fastenings, 98, 156; fencing, 131; filings, 185; fittings, 186, 190; girders, 13, 115, 117, 128; in brickwork, 14; in stonework, 38, 40, 45, 54, 59, 61, 202; laths, 114; piles, 49, 149; plates, 101; principals, 112; rafters, 118; rods, 72; roofs, 67, 69, 72, 150; roof trusses, 142; sashes, 220; skewback, 113; straps, 38, 62, 63; struts, 112; ties, 25, 68; tongue, 66, 177; tubing, 189, 228; turnings, 135; wall ties, 16; wedges, 115; window bars, 128; window frames, 128; work, 36, 42, 60
Italian system of corrugation, 215

JAGGED bolt, 108
Jamb, 35, 38, 62; lining, 176
Jaws, 119, 142, 151
Jog, 52, 136
Joggle, 34, 63, 79, 107; joint, 51, 59, 63, 64, 91, 96, 136
Joggling, 13, 34, 40
Joiners'—stuff, 157, 167, 176; work, 179
Joining, 4
Joint, 46, 61, 65, and *passim*; plates, 142, 152; screw, 164
Jointed or jointer joint, 23
Jointer, 19, 20, 21, 23
Jointing, 4; rule, 19.
Joist, 13, 85, 97
Jump joint, 136
Junction, 7, 8

KEEPING the perpends, 25
Kerbing, 47
Kerbstones, 235
Kerf, 101, 169
Key, 22, 23, 27, 34, 36, 63, 82, 95, 96, 108, 166; joint, 69, 73, 92, 126, 136, 166, 181; stone, 50, 59
Keyed—joint, 23, 86, 92, 166; mitre joint, 166; wedges, 101
Kicks, 30
Kiln bricks, 29
King—bolt, 126; post, 79, 125; rod, 104, 119
King's bond, 15
Knee joint, 195, 228

LABEL, 66
Laid in mouldings, 177
Laminated ribs, 83
Lamp black, 26
Landing, 13, 36, 54, 56, 113
Lap, 68, 70, 71, 74, 77; joint, 23, 69, 105, 128, 192, 194, 233; or lapped dovetail joint, 167; or lapped joint, 52, 92, 136, 166, 195, 215, 218, 221; and mitred dovetail joint, 167; and tongued mitre joint, 167
Lapped mitre joint, 167
Lapping, 68, 91, 93, 104; beams, 98
Larrying, 16
Latchets, 195, 199
Lateral joint, 93, 167
Lathe facing, 116, 131, 153

INDEX.

Lathing, 181
Laths, 67, 68, 75, 114
Laying bricks, 31
Lead, 45, 47, 115; collar, 208; dots, 202, 205; flats, 194, 195; lights, 220, 221; or leaded joint, 52, 93, 136, 195, 221; pegs, 114; plates, 34, 48, 93, 117, 136, 196; plugs, 56, 197; services, 198; spray, 201; strips, 71; tacks, 197; wedges, 58, 198
Leakage, 75, 230
Lean-to roof, 73
Level bed, 27, 51, 57
Levelling, 36, 131
Lewis bolts, 126, 153
Lias lime, 24, 36
Lifting shutters, 176
Light, 26, 27
Lighthouse, 38
Lightning conductor, 45; joint, 137, 218
Lime, 66; and hair plaster, 68; core, 24; putty, 19, 30, 31; stone, 53
Line of pressure, 78, 111
Lining, 33, 166, 173, 176; and jointing, 181; paper, 233
Links, 138, 148
Linseed oil, 53, 59, 84, 122, 142, 179, 229
Lintel, 13; course, 65
Lip, 8, 28, 67; joint, 8, 23; or lipped joint, 94
Lipped joint, 57
Litharge, 53
Lock flooring, 50
Log huts, 79, 109
Longitudinal—bearer, 81; joint, 18, 94, 127, 137, 167, 196; tie, 14
Long joint, 196
Loose joint, 23
Loosening joints, 133
Low pressure pipes, 1:5
Lugs, 67, 118, 124, 132
Luting, 68, 197, 200

MAIN, 147, 150; layers, 200, 228; plates, 120
Marble—dust, 26, 32; work, 45, 47
Margin, 68; drafts, 48, 64
Marini's pipe joint, 139
Marking the courses, 20
Masons' joint, 53
Mastic, 65, 72, 73, 74, 198; joint, 53, 74
Matched boarding, 158
Maul, 34
Medium papers, 233
Meeting surfaces, 100
Mending stone, 42
Messenger's elastic joint, 193
Metal joint, 4
Meter, 229
Mitre, 157; and butt joint, 168; block, 168; box, 168; clamped joint, 168; joint, 23, 53, 69, 137, 167, 182, 196, 215, 216, 233; joint keyed, 168; shoot, 168
Mitred—and cross tongued joint, 168; dovetail joint, 168

Mitring slates, 73
Moisture, 80, 88, 89, 94
Molten lead, 58, 137
Mortar, 13, 18, 54; bed, 25; fillet, 20; making, 24; joint, 24, 25, 27, 32, 53, 54, 58, 69, 74
Mortise, 89, 92, 95, 117; and tenon, 81, 84, 88, 103; and tenon joint, 54, 95, 137, 169; holes, 46; joint, 90, 95
Mouldings, 53, 55, 60, 61, 65
Movable joints, 119, 173
Moving loads, 144
Mullions, 52, 56, 62
Multiple riveting, 153
Musgrave's adamantine clinkers, 22

NAILED joint, 169
Nail hold, 169
Nailing slates, 71, 74
Naphtha, 42
Neat cement, 16, 42
Needles, 78
Neutral portion, 95
New work, 14, 63
Niches, 57
Nipple unions, 232
Nose—bit, 231; piece, 228
Notch, 14, 84, 85, 88, 95, 97, 102, 136; and bridle joint, 97; or notched joint, 54, 97
Notched, 63; joint, 169
Notching, 79, 87, 90, 92, 103
Nuts, 82, 93, 150; and washers, 106

OAK—dowels, 68; laths, 70; pins, 67, 69; templates, 80; treenails, 108
Oakum, 121
Oblique—angles, 15; joint, 97, 98
Obliquity, 98, 168
Octagonal sockets, 149
Oil, 26; cement, 72, 73; mastic, 65; putty, 73, 74
Old—English bond, 15; work, 25, 63
Openings, 15, 16, 41
Open joint, 21, 52, 55, 65, 74, 98, 142, 170, 172, 222
Ordinary stocks, 30
Ornamental—castings, 116; ironwork, 137; tiles, 67, 68
Out of winding, 36, 47, 62, 64
Outgo of trap, 206
Overcast — joint, 196; ribbon joint, 196
Overhead girders, 115, 116, 117, 124, 132, 148
Overhang, 22
Overlapping joint, 196, 222
Overrunning the courses, 25
Oversailing, 20; courses, 22
Ox hair, 67, 68
Oxidation, 42, 84, 107, 109, 144, 145, 156, 202

PACKING, 8, 11, 40, 44, 80, 109, 114, 115, 133, 152, 204

Painted joint, 72, 73, 74, 75, 222
Painter's pipe joint, 130
Painting, 53
Pallets, 32
Panel, 221
Panelled walls, 23
Panelling, 158, 181, 233
Pantiles, 68
Parallel — joint, 25; surfaces, 153; tenons, 26
Parapet, 23
Partitions, 80, 83, 96, 106
Patent axe, 48
Pat joint, 12, 25
Paving, 22, 23, 54, 234
Pebbles, 52, 64
Pediments, 55, 59
Pencil joint, 233
Pendants, 229, 231
Percussive action, 41
Perpendicular joint, 25, 55
Perpends, 18, 65
Pick, 48
Picked—masonry, 36; stocks, 29
Piecing, 27, 30; joint, 25, 98, 170
Pier, 23, 26, 46, 61, 111, 114, 130; walls, 50
Pilasters, 13, 115, 122, 181
Pile, 104, 119, 124, 126, 134; foundations, 109
Piling, 57, 120, 124, 153, 154, 155
Pillars, 126, 132
Pin, 46, 87, 95, 98, 112, 127, 130, 148; hole, 138, 200; joint, 138
Pinned joint, 55, 98, 170
Pinning, 36, 64; in, 55
Pipe—joint, 8, 138, 150, 197, 228; nails, 139; socket, 85
Piping, 121, 133, 150, 153
Pitch, 60, 68, 69, 85, 234, 236; lines, 128, 156; of rivets, 122, 128, 153; of roof, 69, 70, 76; of screw, 82
Pitching, 235
Pivot, 101; bridge, 140; lights, 140; or pivoted joint, 139
Pivoted centres, 140
Pivoting, 154
Plain—ends, 150; or plane joint, 100, 170; tiles, 67, 69
Planed—bed plates, 117; joint, 140; surfaces, 111
Planing, 130, 131; machine, 123, 140
Planking, 85; and strutting, 87, 93; foundations, 87
Planks, 104
Planted in mouldings, 168
Plaster, 135; of Paris, 42, 45
Plasterers' putty, 48, 54
Plastering, 26, 109, 182;
Plate, 13, 52, 62, 81, 104, 109; iron, 140, 142
Platform, 85, 110
Plough, 163, 171
Ploughed — and feathered joint, 171; and tongued joint, 171
Plugged joint, 56, 171, 197

Plugging, 64, 171
Plugs, 34, 46, 56, 60, 150, 197, 230
Plumbago, 173
Plumbers'—cement, 188; damage, 77 solder, 205
Pockets, 41
Pointed—joint, 26, 56, 69, 75; masonry, 36; sashes, 159; work, 20
Pointer, 23
Pointing, 12, 17, 26, 28, 37, 58, 65, 198; stuff, 12, 20; tiles, 69
Polished joint, 59
Porosity, 7, 10, 66
Portland cement, 6, 7, 16, 21, 36, 37, 41, 42, 45, 50, 53, 56, 59, 60, 65, 66, 68, 73, 115
Post, 78, 96, 97, 102, 104, 109; and beam joint, 100; stones, 36, 136, 140
Pot metal, 205, 210
Pouring stick, 208
Pressed lead pipe, 196, 203
Pressure, 13, 28, 34, 52, 93, 98, 231
Principal, 102, 112, 129; or principal rafter, 78, 79, 84, 93, 96, 97, 103, 105
Profile, 51
Proper door frame, 100
Protuberances, 43, 48, 123
Puddled clay, 7
Puddling, 7
Pulley mortise joint, 100
Pump—joint, 133; valve, 188
Punch, 48, 141
Punched joint, 140
Puncheons, 109
Punching, 132, 140, 141, 144; machine, 129
Purlins, 97, 103, 114, 119, 143
Putty, 11, 26, 43, 48, 119, 139, 222; joint, 26, 57, 198, 222
Puttyless joint, 223

QUARRIES, 22
Quarters, 109
Quay wall, 46
Queen—post, 103; rod, 104, 119
Quirked bead, 157
Quoins, 15, 16, 53, 60, 63

RABBETED joint, 171
Racked—joint, 236; back, 25, 26
Racking, 2, 83, 96
Radial—joint, 141; motion, 123
Radiating—bed joint, 56; joint, 26, 57, 100, 114, 141
Radius joint, 171
Rafter, 91, 100, 104, 106, 114, 117, 119, 150
Raglet, 58, 198, 202, 205; joint, 57, 199
Rail, 177
Railings, 54
Rain, 18, 27, 61, 65, 176, 178; water pipes, 139, 184, 204
Raised joint, 26, 58
Raked joint, 26, 58
Raker, 78

INDEX. 245

Raking—bed, 51; bed joint, 56; bond, 15, 19; out, 26, 56, 58; strut, 79
Rammer, 9
Random rubble, 62
Range of temperature, 129
Ranging bond, 33, 169
Ratchet brace, 132
Rebate, 55, 58, 62; joint, 73; plane, 164, 172
Rebated—and beaded joint, 172; and filleted joint, 172; and mitred joint, 172; grooved and tongued joint, 172; joint, 58, 75, 100, 141, 171
Rebating, 60; and bolting, 90
Recessed joint, 26, 58
Red—cement, 66, 72, 133; heat, 156; lead, 119, 121, 122, 139, 154, 155; lead joint, 142; putty, 26
Reduced brickwork, 16, 24
Relieving arch, 11
Rendering, 26, 27, 75, 182
Rendle's system of glazing, 223
Repointing, 18
Reservoir walls, 42
Resilience, 171
Resin, 42, 186, 187, 188, 201, 220, 227
Resultant pressure, 39, 134
Retaining walls, 113
Rib, 132, 141
Ribbed groins, 57
Ribbon joint, 199
Ridge, 12, 61, 69, 71, 72, 74, 195; piece, 73, 100, 104; roll, 67; tiles, 68
Ridging, 53, 71, 72, 73, 75
Rigidity, 151
Rimer, 132, 141
Ring, 11, 124; course, 18, 22, 50
Rising course, 14
Rivet, 13, 114, 127, 144; head, 141; holes, 129, 141; iron, 142
Riveted, 105; joint, 122, 142
Riveting, 131, 141, 144; machine, 143; set, 143
Roadway planking, 234
Rockworked quoins, 60
Rod, 114, 148; bolts, 38, 49, 62; of brickwork, 16, 24
Rodding, 135
Roll, 65, 68; joint, 199, 215, 219
Rolled—iron, 133; joint, 200, 233
Roller, 112, 117, 130, 134, 140; frame, 130, 146; path, 140
Roman cement, 16, 42
Roof, 98, 103; principals, 129
Root of a tenon, 95
Rough—arch, 11; boarding, 77, 186; joint, 26; stocks, 29; stuff, 181; tooled joint, 59
Rounded—ends, 128; joint, 86, 101, 173; off, 86
Round joint, 200
Rubbed—joint, 27, 59; work, 19
Rubbing, 12, 27, 62; down, 30; stone, 27
Rubbish, 16, 24
Rubble, 40, 41, 44, 49, 65; masonry, 39, 54, 58; walling, 54, 56, 63

Rule joint, 85, 171, 173
Run joint, 52, 59, 145, 200
Runner, 140, 173, 177
Running—bond, 15; joint, 173; mouldings, 182; nut, 192; sand, 7; water, 42
Rushes, 227
Rust, 119, 135, 228; cement, 133; joint, 145
Rustic—joint, 58; or rusticated joint, 60, 62; work, 43, 181
Rustics, 60

SADDLE, 61, 75, 138; back fillet, 178; cap, 213; joint, 27, 60, 146, 215, 228; piece, 27, 61, 215; stone, 55
Saddling—a main, 146; a service, 229; the joint, 60, 65, 66
Sal ammoniac, 146, 189
Sally, 85
Salt, 201
Sand, 16; joint, 146
Sanded and tarred hoop iron, 16
Sarking, 76
Sash, 177; bar, 162, 168, 175, 222; door, 224; panes or squares, 222; window, 161, 176
Sawn—joint, 61; slate, 72
Scaffold cords, 110
Scantling, 101
Scarf, 91, 98, 101, 104, 108; or scarfed joint, 101, 173
Scarfed beam, 83
Scraping, 186, 195; down, 123
Scratching, 182
Screeded, 22
Screw, 98, 146; bolt, 81, 106, 114, 118; cramp, 159; ends, 62, 148; ferrule, 132; joint, 146, 174, 202, 215, 229; nail, 147, 174; pile, 119, 136, 141; shackle, 126; shackle joint, 102, 147; thread, 82, 112, 132
Scribed joint, 61, 174
Scribing, 174
Scoring, 182
Seam, 85, 94; joint, 175, 202
Seasoned timber, 80, 94, 96, 170
Seat, 88, 113
Seating, 112, 115
Sea wall, 49
Secret—dovetail, 168; dovetail joint, 175; nailed joint, 175
Serrated—joint, 102; tabulations, 103, 109
Service pipes, 139, 198
Sets, 234, 235
Setting—of mortar, 25, 66; up lead, 188, 198, 200
Settlement, 3, 7, 8, 9, 13, 18, 25, 27, 74, 79, 80, 85, 107, 188, 231
Shackle joint, 148
Shaft, 55, 115, 202
Shakiness, 13
Sham joint, 30
Shank, 143
Shavehook, 205

Shearing, 98, 99, 128, 129, 138, 141, 143, 144, 147
Sheet piling, 49, 79, 85, 89, 124, 154
She joggle, 43, 50
Shellac, 42
Shelving, 160
Shift joint, 27, 61
Ship lapping joint, 103
Shocks, 44
Shoe, 103, 117, 118, 126, 148 ; joint, 148, 175 ; plate, 80
Shop window, 220, 221, 223
Shores, 78, 104
Short joint, 203
Shot edges, 158, 161, 162, 178
Shoulder, 75, 132 ; of a tenon, 95
Shouldered joint, 75, 103
Shouldering, 75, 92, 103
Shrinkage, 3, 18, 29, 79, 80, 85, 88, 96, 107, 159, 164, 166, 172
Shutter, 177
Shutting joint, 158, 176
Side—fillister, 161, 164, 172 ; joint, 46, 47, 56, 61, 64, 65, 75
Sill, 65, 79, 105, 109, 166, 176, 178, 222 ; course, 65
Silver—sand, 26, 31, 32 ; solder, 217
Simple joint, 2
Single—herringbone paving, 19 ; riveted joint, 148
Sink, 58 ; stone, 58
Sinking, 13, 60, 161, 175, 199, 222
Size, 26
S-joint, 146, 215, 223
Skew—arch, 35, 47, 57 ; back, 11, 61, 113 ; quoin, 14, 15 ; tile, 69
Skin of castings, 145
Skirting, 166, 168, 170, 175
Skylight, 119, 179, 194, 221
Slab, 8, 49, 54
Slate, 114, 194 ; cramp, 45 ; dowel, 47 ; mastic, 72, 74 ; or slating nails, 71 ; roll, 75 ; slab, 75
Sleepers, 13
Sleeve, 114, 126
Sliding, 56, 97, 100, 102, 103, 113, 130, 133, 136 ; joint, 176, 229
Slip, 85, 121, 133, 169 ; feather, 161, 166, 178 ; feather joint, 177 ; joint, 25, 27, 177, 203 ; socket joint, 204
Slipping, 106, 116, 125
Sloping bed, 51
Slot, 13, 83, 107, 114, 115, 138
Smiths' ashes, 14, 74
Smooth joint, 62
Smoothed joint, 44
Smudge, 210
Snap, 143 ; rivet, 143
Snapped header, 15, 29
Snatching, 181
Snecked walling, 62
Snow, 70, 75, 76
Snug, 13, 116, 117
Soaker, 73
Soaking bricks, 29
Socket, 8, 9, 80, 88, 104, 113, 114, 116, 124, 126, 229 ; joint, 9, 103, 149, 192, 204, 229 ; or socket piece, 103
Soffit, 11, 19, 34, 57, 181
Soft solder, 205
Soil, 210 ; pipe, 196, 197, 204, 206
Soiling, 195
Solder, 133, 205
Soldered joint, 149, 205, 215, 229
Soldering—cloth, 209 ; iron, 189 ; tool, 189
Sole plate, 145
Solid work, 21, 28
Spacing rivets, 121
Spalls, 37, 48
Span, 142 ; roof, 73
Spandril, 35 ; arch, 113
Spanish brown, 26
Spelter solder, 217
Spherical work, 43
Spigot, 8, 9, 125 ; and faucet joint, 150 and socket joint, 150
Spiked joint, 104
Spikes, 101, 104
Spindle, 81, 82, 143
Spire ashlar, 51, 57
Spirelet, 83, 106
Spirit of salt, 190, 215
Splash board, 179
Splashing stick, 208
Splayed joint, 104
Splice joint, 104, 165, 177
Split, 21, 40, 202 ; key, 138
Splitting, 102, 103, 109
Spray, 187
Spreading, 105, 108
Sprigged joint, 177, 223
Springed or sprung boarding, 93
Springer, 61
Springing, 34, 47, 113 ; motion, 123, 130 ; of plates, 153 ; pipes, 7
Spring joint, 27, 177
Spun yarn, 85, 121
Square—edged joint, 167, 170 ; halving, 90 ; joint. 39, 62, 104, 178 ; piers, 114
Squint quoin, 14
Staff rivet, 144
Stains in stonework, 42, 60
Stairs, 177, 180
Stanchions, 54, 114, 116
Standard, 54, 119, 122, 131
Stanford joint, 7, 9
Stay bar, 126
Steel—bolt, 118 ; core, 188
Step, 55
Stepped flashing, 194
Stick of lead, 184, 188
Stiffeners, 115, 119, 132, 142
Stiff—joint, 229, 231 ; mortar, 18, 22
Stiffness, 124
Stiles, 158, 165, 176
Stirrup, 80, 105, 106 ; joint, 105
Stocks, 29, 30
Stone—jambs, 38 ; paving, 235 ; sill, 66 ; steps, 51, 52, 57 ; template, 38, 80, 130 ; windows, 220, 222
Stoneware pipes, 9, 185

INDEX. 247

Stop, 221, 230; ends, 213, 218
Stopping, 26, 27, 30, 56, 58
Straight—arch, 11; edge, 20, 21; joint, 27, 36, 62, 178
Strain, 5, 78
Straining, 101; beam, 103, 104; piece, 79, 107
Strap, 89, 104, 138; and bolt, 102; and bolt joint, 106; and cotter joint, 107; or strapped joint, 62, 105
Stretchers, 15, 35;
Striking—centres, 34, 110; platforms, 110; the joints, 27
String, 53, 65, 66, 166, 175, 180; course, 18, 50
Strip, 52; lead, 73
Striped joint, 206
Struck—joint, 20, 27; or stuck bead, 164, 176; or stuck moulding, 177
Strut, 79, 88, 97, 104, 105, 112, 116; joint, 84, 107, 150
Strutting to floors, 102, 148
Stub, 46, 68, 117, 122, 125, 137, 175; tenon, 92, 100
Stucco, 22, 181
Stud, 109, 114
Stuffing, 8, 121, 229
Submerged—foundations, 155; timbers, 90
Subsidence, 6
Subsoil—drain-pipes, 7; water, 8
Sulphur, 9, 45, 47, 115, 145, 146, 151; joint, 151
Summering, 57
Sunk—joint, 28, 62; stone, 117
Superimposed columns, 116, 117, 149, 155
Surface—drain, 22; pipes, 7
Surfacing iron, 123, 140
Suspending—piece, 102, 106; rod, 127, 148
Swallow-tail joint, 107
Swelling of—cement, 18; lime, 22
Swing—bridge, 140; joint, 229
Swivel joint, 206, 230
Syphonic action, 44

TABLE, 62
Tabled—indented and keyed joint, 108; joint, 62, 108
Tabling, 101
Tacks, lead, 204
Tafting, 207
Taft joint, 206, 210
Tail, 40, 64, 68, 74, 106, 138, 143
Tallow, 154, 188, 209, 220, 223, 227, 230
Tambour, 37, 48
Tanpin, 186
Tapping—main, 131; rivets, 143
Tar, 9, 61, 155, 234
Tarnish, 210
Tarred—gasket, 133; hoop iron, 16
Tearing, 128, 144
Template, 13, 36, 80, 113
Templet, 101, 143
Tenon 54, 89, 95

Tensile resistance or strength, 98, 138; stress, 89, 102, 103, 108, 121
Tension, 108, 148, 188; bar, 138; rod, 126, 147, 148
Terra cotta, 17, 23; joint, 28
Tesseræ, 22
Thermo-plastic putty joint, 223
Thick joint, 18, 25, 29, 63
Thimble, 152; joint, 152
Thin joint, 11, 17, 18, 28, 29, 63
Thread, 55, 65; of screw, 82, 147
Throating, 66, 178, 179
Through — and through joint, 152; stones, 40
Tie, 113, 124, 131; bar, 62, 122, 152; beam, 78, 79, 80, 84, 87, 96, 105, 106, 125; bolt, 120; joint, 30, 63, 84, 152; plate, 80; rod, 102, 113, 119, 124, 127, 142, 150
Tight joint, 9, 10, 68, 89, 108, 135, 152, 153, 193, 207, 230
Tile, 22; pavement, 234; pins and nails, 67
Tilting—action, 113; fillet, 71
Timber, 13, 104; bridge, 93, 94, 98; floor, 57; joints, 96; partition, 152
Tirgles, 195, 199, 221
Tinning, 189, 208, 227
Tin pan, 189
T—iron, 124, 142, 150; joint, 151, 206, 229, 230; nails, 67, 68 69
Tompion, 186
Tongue, 68, 84, 97, 159, 176, 177; and groove, 52
Tongued, 89; joint, 182; notch, 84
Ton slates, 76
Tool, 48
Tooled joint, 63
Tooling, 35, 62, 63
Toothed—joint, 30, 109, 182; work, 25
Toothing, 30, 101
Top coating, 236
Torched joint, 76
Torching, 71, 76
Torsion, 136
Touch, plumbers', 203, 205, 210
T-piece, 125, 139, 151
Tracery, 41, 56, 57
Transom, 54, 62
Transverse—joint, 14, 18, 30, 64, 65, 109; strain, 101, 108
Traps, 62
Treenail, 95, 99
Trimmer, 99, 109, 110
Trimming joist, 99, 109
Triple—chain riveted lap joint, 123; riveted joint, 153
Trowel, 56; crease, 73
True—bearing, 100; joint, 64, 153; pitch, 70, 78
Truss, 79, 83, 87, 96, 98, 104, 106, 113, 130, 152
Trussed girder, 83
Tubing, 228
Tuck—and pat joint, 12; pointed joint, 30; pointing, 20, 26

INDEX.

Turned and bored join*, 153
Turnings, iron, 146
Turnpin, 186, 189
Twist, 48, 62. 64
Twisting, 170 ; strain, 127
Tyerman's hoop iron, 16

UNBAKED bricks, 169
Uncoursed—ashlar, 39 ; rubble, 44
Undercutting, 66
Underground work, 18
Underhand joint, 207, 211
Underpinned joint, 64
Uniform—bearing, 59 ; strength, 149
Union, 126, 232 ; joint, 154, 208, 232
Universal—joint, 154, 232 ; swivel, 232
Upright, 132 ; joint, 18, 31, 64, 208

VACUITIES, 40
Vacuum, 10
Valley, 71, 195 ; tiles, 67
Vascular plates, 98
Vegetable black, 26
Ventilated—pan tiling, 69 ; slating, 74
Ventilating pipes, 193
Ventilation, 10 ; of soil pipes, 10, 185
Ventilators, 72, 119
Vent linings, 41
Verandah floors, 197, 209
Vertical joint, 14, 18, 25, 55, 60, 61, 64, 65, 76, 154 ; system of breakwater construction, 58
Vibration, 5, 116, 144
V joint, 53, 64, 178
V M zinc nails, 77
Voids, 40, 44, 50
Voussoirs, 34, 47, 48, 52, 111, 114

WAINSCOTING, 33, 87
Walings, 104
Wall—hooks, 194 ; joint, 31 ; plates, 78, 85, 97, 100, 101, 107 ; tiles, 68.
Warehouses, 132
Washers, 38, 83, 101, 106, 107, 109, 133, 151, 226, 228
Wash leather joint, 224
Waste pipes, 197, 204
Water, 21, 24, 228 ; and lime, 50 ; bar, 177, 178 ; fillet, 179 ; joint, 65, 232 ; lock, 228 ; mains, 125

Water-tight, 85, 122 ; joint, 9, 109, 154, 178
Wave action, 90
Weather, 98 ; board, 179 ; joint, 32, 65, 69, 76, 178 ; slating, 75 ; tiling, 67, 68
Weathering, 21, 61, 65, 66, 68, 98
Weather-tight, 75, 194 ; joint, 14
Webs, 111
Wedged joint, 109, 155, 179, 209, 216
Wedges, 52, 58, 64, 85, 88, 92, 102, 109, 149, 169, 180, 209
Weeping, 206
Weirs, 44
Welded joint, 155, 197, 209
Welding, 105, 126 ; heat, 137
Welds, 112
Welted joint, 216
Wet, 28, 32 ; bricks, 22, 25
Wetting brickwork, 22 ; work, 26, 65, 73
White—heat, 149, 155 ; lead, 48, 49, 59, 121, 122, 139, 142, 154, 163, 176, 179 ; lead joint, 66 ; lime putty, 12 ; joint, 12, 32, 75 ; putty, 26
Whiting, 154
Wind, 32, 65, 76, 144, 176 ; bracing, 119
Winders, 110
Window—bar, 62, 128 ; frame, 13, 32, 128 ; head, 51, 221 ; lead, 164 ; lining, 166 ; plane, 62 ; sills, 23, 51, 53
Wiped joint, 186, 209
Wiping, 210 ; cloth, 209 ; solder, 186
Withs, 30
Wood—blocks, 185, 236 ; bricks, 13, 169 ; fillet, 133 ; frames, 198 ; joint, 32 ; paving, 234, 235, 236 ; pin, 99 ; plug, 171 ; roll, 195, 199 ; screw, 147, 174, 202 ; sill, 66, 178, 179 ; slip, 32, 133, 169
Working—centre, 57 ; strength, 83
Wrought iron piping, 137

YARD paving, 236
Yarn, 139, 150
York paving, 235

ZINC, 60 ; bead, 214 ; clips, 114, 214 ; roll cap joint, 216

THE END.

PRINTED BY J. S. VIRTUE AND CO. LIMITED CITY ROAD LONDON.

Weale's Rudimentary Series.

PHILADELPHIA, 1876.
THE PRIZE MEDAL
Was awarded to the Publishers for
Books: Rudimentary, Scientific,
"WEALE'S SERIES," ETC.

A NEW LIST OF
WEALE'S SERIES
RUDIMENTARY SCIENTIFIC, EDUCATIONAL, AND CLASSICAL.

Comprising nearly <u>Three Hundred and Fifty</u> distinct works in almost every department of Science, Art, and Education, recommended to the notice of <u>Engineers, Architects, Builders, Artisans, and Students generally</u>, as well as to those interested in <u>Workmen's Libraries, Literary and Scientific Institutions, Colleges, Schools, Science Classes</u>, &c., &c.

☞ "WEALE'S SERIES includes Text-Books on almost every branch of Science and Industry, comprising such subjects as Agriculture, Architecture and Building, Civil Engineering, Fine Arts, Mechanics and Mechanical Engineering, Physical and Chemical Science, and many miscellaneous Treatises. The whole are constantly undergoing revision, and new editions, brought up to the latest discoveries in scientific research, are constantly issued. The prices at which they are sold are as low as their excellence is assured."—*American Literary Gazette.*

"Amongst the literature of technical education, WEALE'S SERIES has ever enjoyed a high reputation, and the additions being made by Messrs. CROSBY LOCKWOOD & Co. render the series even more complete, and bring the information upon the several subjects down to the present time."—*Mining Journal.*

"It is impossible to do otherwise than bear testimony to the value of WEALE'S SERIES."—*Engineer.*

"Everybody—even that outrageous nuisance 'Every Schoolboy'—knows the merits of 'WEALE'S RUDIMENTARY SERIES.' Any persons wishing to acquire knowledge cannot do better than look through Weale's Series and get all the books they require. The Series is indeed an inexhaustible mine of literary wealth."—*The Metropolitan.*

"WEALE'S SERIES has become a standard as well as an unrivalled collection of treatises in all branches of art and science."—*Public Opinion.*

LONDON, 1862.
THE PRIZE MEDAL
Was awarded to the Publishers of
"WEALE'S SERIES."

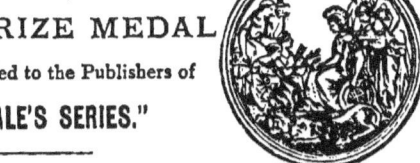

CROSBY LOCKWOOD & CO.,
7, STATIONERS' HALL COURT, LUDGATE HILL, LONDON, E.C.

WEALE'S RUDIMENTARY SCIENTIFIC SERIES.

※※ The volumes of this Series are freely Illustrated with Woodcuts, or otherwise, where requisite. Throughout the following List it must be understood that the books are bound in limp cloth, unless otherwise stated; *but the volumes marked with a ‡ may also be had strongly bound in cloth boards for 6d. extra.*

N.B.—*In ordering from this List it is recommended, as a means of facilitating business and obviating error, to quote the numbers affixed to the volumes, as well as the titles and prices.*

ARCHITECTURE, BUILDING, ETC.

No.
16. *ARCHITECTURE—ORDERS*—The Orders and their Æsthetic Principles. By W. H. LEEDS. Illustrated. 1s. 6d.
17. *ARCHITECTURE—STYLES*—The History and Description of the Styles of Architecture of Various Countries, from the Earliest to the Present Period. By T. TALBOT BURY, F.R.I.B.A., &c. Illustrated. 2s.
   ※※ ORDERS AND STYLES OF ARCHITECTURE, *in One Vol.*, 3s. 6d.
18. *ARCHITECTURE—DESIGN*—The Principles of Design in Architecture, as deducible from Nature and exemplified in the Works of the Greek and Gothic Architects. By E. L. GARBETT, Architect. Illustrated. 2s.6d.
※※ *The three preceding Works, in One handsome Vol., half bound, entitled* "MODERN ARCHITECTURE," *price 6s.*
22. *THE ART OF BUILDING*, Rudiments of. General Principles of Construction, Materials used in Building, Strength and Use of Materials, Working Drawings, Specifications, and Estimates. By E. DOBSON, 2s.‡
23. *BRICKS AND TILES*, Rudimentary Treatise on the Manufacture of; containing an Outline of the Principles of Brickmaking. By EDW. DOBSON, M.R.I.B.A. With Additions by C. TOMLINSON, F.R.S. Illustrated, 3s.‡
25. *MASONRY AND STONECUTTING;* in which the Principles of Masonic Projection and their application to the Construction of Curved Wing-Walls, Domes, Oblique Bridges, and Roman and Gothic Vaulting, are explained. By EDWARD DOBSON, M.R.I.B.A., &c. 2s. 6d.‡
44. *FOUNDATIONS AND CONCRETE WORKS*, a Rudimentary Treatise on; containing a Synopsis of the principal cases of Foundation Works, with the usual Modes of Treatment, and Practical Remarks on Footings, Planking, Sand, Concrete, Béton, Pile-driving, Caissons, and Cofferdams. By E. DOBSON, M.R.I.B.A., &c. Fifth Edition, revised. 1s. 6d.
42. *COTTAGE BUILDING.* By C. BRUCE ALLEN, Architect. Ninth Edition, revised and enlarged. Numerous Illustrations. 1s. 6d.
45. *LIMES, CEMENTS, MORTARS, CONCRETES, MASTICS*, PLASTERING, &c. By G. R. BURNELL, C.E. Twelfth Edition. 1s. 6d.
57. *WARMING AND VENTILATION*, a Rudimentary Treatise on; being a concise Exposition of the General Principles of the Art of Warming and Ventilating Domestic and Public Buildings, Mines, Lighthouses, Ships, &c. By CHARLES TOMLINSON, F.R.S., &c. Illustrated. 3s.
83**. *CONSTRUCTION OF DOOR LOCKS.* Compiled from the Papers of A. C. HOBBS, Esq., of New York, and Edited by CHARLES TOMLINSON, F.R.S. To which is added, a Description of Fenby's Patent Locks, and a Note upon IRON SAFES by ROBERT MALLET, M.I.C.E. Illus. 2s. 6d.
111. *ARCHES, PIERS, BUTTRESSES, &c.:* Experimental Essays on the Principles of Construction in; made with a view to their being useful to the Practical Builder. By WILLIAM BLAND. Illustrated. 1s. 6d.
116. *THE ACOUSTICS OF PUBLIC BUILDINGS;* or, The Principles of the Science of Sound applied to the purposes of the Architect and Builder. By T. ROGER SMITH, M.R.I.B.A., Architect. Illustrated. 1s. 6d.

☞ *The ‡ indicates that these vols. may be had strongly bound at 6d. extra.*

LONDON : CROSBY LOCKWOOD AND CO.,

Architecture, Building, etc., *continued.*

127. *ARCHITECTURAL MODELLING IN PAPER,* the Art of. By T. A. RICHARDSON, Architect. Illustrated. 1s. 6d.

128. *VITRUVIUS—THE ARCHITECTURE OF MARCUS VITRUVIUS POLLO.* In Ten Books. Translated from the Latin by JOSEPH GWILT, F.S.A., F.R.A.S. With 23 Plates. 5s.

130. *GRECIAN ARCHITECTURE,* An Inquiry into the Principles of Beauty in; with an Historical View of the Rise and Progress of the Art in Greece. By the EARL OF ABERDEEN. 1s.

*** *The two preceding Works in One handsome Vol., half bound, entitled* "ANCIENT ARCHITECTURE," *price* 6s.

132. *DWELLING-HOUSES,* a Rudimentary Treatise on the Erection of.‡ Illustrated by a Perspective View, Plans, Elevations, and Sections of a pair of Semi-detached Villas, with the Specification, Quantities, and Estimates, and every requisite detail, in sequence, for their Construction and Finishing. By S. H. BROOKS, Architect. New Edition, with Plates. 2s. 6d.‡

156. *QUANTITIES AND MEASUREMENTS,* How to Calculate and Take them in Bricklayers', Masons', Plasterers', Plumbers', Painters', Paperhangers', Gilders', Smiths', Carpenters', and Joiners' Work. By A. C. BEATON, Architect and Surveyor. New and Enlarged Edition. Illus. 1s. 6d.

175. *LOCKWOOD & CO.'S BUILDER'S AND CONTRACTOR'S* PRICE BOOK, for 1882, containing the latest Prices of all kinds of Builders' Materials and Labour, and of all Trades connected with Building, &c., &c. Revised and Edited by F. T. W. MILLER, Architect and Surveyor. 3s. 6d.; half bound, 4s.

182. *CARPENTRY AND JOINERY—*THE ELEMENTARY PRINCIPLES OF CARPENTRY. Chiefly composed from the Standard Work of THOMAS TREDGOLD, C.E. With Additions from the Works of the most Recent Authorities, and a TREATISE ON JOINERY by E. WYNDHAM TARN, M.A. Numerous Illustrations. 3s. 6d.‡

182*. *CARPENTRY AND JOINERY. ATLAS* of 35 Plates to accompany the above. With Descriptive Letterpress. 4to. 6s.; cloth boards, 7s. 6d.

187. *HINTS TO YOUNG ARCHITECTS.* By GEORGE WIGHTWICK. New, Revised, and enlarged Edition. By G. HUSKISSON GUILLAUME, Architect. With numerous Woodcuts. 3s. 6d.‡

188. *HOUSE PAINTING, GRAINING, MARBLING, AND SIGN WRITING:* A Practical Manual of, containing full information on the Processes of House-Painting, the Formation of Letters and Practice of Sign-Writing, the Principles of Decorative Art, a Course of Elementary Drawing for House-Painters, Writers, &c., &c. With 9 Coloured Plates of Woods and Marbles, and nearly 150 Wood Engravings. By ELLIS A. DAVIDSON. Third Edition, revised. 5s. cloth limp; 6s. cloth boards.

189. *THE RUDIMENTS OF PRACTICAL BRICKLAYING.* In Six Sections: General Principles; Arch Drawing, Cutting, and Setting; Pointing; Paving, Tiling, Materials; Slating and Plastering; Practical Geometry, Mensuration, &c. By ADAM HAMMOND. Illustrated. 1s. 6d.

191. *PLUMBING.* A Text-Book to the Practice of the Art or Craft of the Plumber. With Chapters upon House Drainage, embodying the latest Improvements. Third Edition, enlarged. Containing 300 Illustrations. By W. P. BUCHAN, Sanitary Engineer. 3s. 6d.‡

192. *THE TIMBER IMPORTER'S, TIMBER MERCHANT'S,* and BUILDER'S STANDARD GUIDE; comprising copious and valuable Memoranda for the Retailer and Builder. By RICHARD E. GRANDY. Second Edition, Revised. 3s.‡

205. *THE ART OF LETTER PAINTING MADE EASY.* By J. G. BADENOCH. Illustrated with 12 full-page Engravings of Examples. 1s.

206. *A BOOK ON BUILDING, Civil and Ecclesiastical,* including CHURCH RESTORATION. With the Theory of Domes and the Great Pyramid, &c. By Sir EDMUND BECKETT, Bart., LL.D., Q.C., F.R.A.S. Second Edition, enlarged, 4s. 6d.‡

☞ *The* ‡ *indicates that these vols. may be had strongly bound at* 6d. *extra.*

7, STATIONERS' HALL COURT, LUDGATE HILL, E.C.

Architecture, Building, etc., *continued.*

226. **THE JOINTS MADE AND USED BY BUILDERS** in the Construction of various kinds of Engineering and Architectural Works (A Practical Treatise on). With especial reference to those wrought by Artificers in Erecting and Finishing Habitable Structures. By WYVILL J. CHRISTY, Architect and Surveyor. With upwards of One Hundred and Sixty Engravings on Wood. 3s.‡ [*Just published.*

228. **THE CONSTRUCTION OF ROOFS OF WOOD AND IRON** (An Elementary Treatise on). Deduced chiefly from the Works of Robison, Tredgold, and Humber. By E. WYNDHAM TARN, M.A., Architect. With numerous Illustrations. 1s. 6d. [*Just published.*

229. **ELEMENTARY DECORATION:** A Guide to the Simpler Forms of Everyday Art, as applied to the Interior and Exterior Decoration of Dwelling-Houses, &c. By JAMES W. FACEY, Jun. Illustrated with Sixty-eight explanatory Engravings, principally from Designs by the Author. 2s. [*Just published.*

230. **HANDRAILING** (A Practical Treatise on). Showing New and Simple Methods for finding the Pitch of the Plank, Drawing the Moulds, Bevelling, Jointing-up, and Squaring the Wreath. By GEORGE COLLINGS. Illustrated with Plates and Diagrams. 1s. 6d. [*Just published.*

CIVIL ENGINEERING, ETC.

219. **CIVIL ENGINEERING.** By HENRY LAW, M.Inst. C.E. Including a Treatise on HYDRAULIC ENGINEERING by GEO. R. BURNELL, M.Inst.C.E. Sixth Edition, revised, WITH LARGE ADDITIONS ON RECENT PRACTICE IN CIVIL ENGINEERING, by D. KINNEAR CLARK, M.Inst. C.E., Author of "Tramways: Their Construction," &c. 6s. 6d., Cloth boards, 7s. 6d.

31. **WELL-DIGGING, BORING, AND PUMP-WORK.** By JOHN GEORGE SWINDELL, A.R.I.B.A. New Edition, by G. R. BURNELL, C.E. 1s. 6d.

35. **THE BLASTING AND QUARRYING OF STONE**, for Building and other Purposes. With Remarks on the Blowing up of Bridges. By Gen. Sir JOHN BURGOYNE, Bart., K.C.B. Illustrated. 1s. 6d.

62. **RAILWAY CONSTRUCTION**, Elementary and Practical Instructions on the Science of. By Sir M. STEPHENSON, C.E. New Edition, by EDWARD NUGENT, C.E. With Statistics of the Capital, Dividends, and Working of Railways in the United Kingdom. By E. D. CHATTAWAY. 4s.

80*. **EMBANKING LANDS FROM THE SEA**, the Practice of. Treated as a Means of Profitable Employment for Capital. With Examples and Particulars of actual Embankments, &c. By J. WIGGINS, F.G.S. 2s.

81. **WATER WORKS**, for the Supply of Cities and Towns. With a Description of the Principal Geological Formations of England as influencing Supplies of Water; and Details of Engines and Pumping Machinery for raising Water. By SAMUEL HUGHES, F.G.S., C.E. New Edition. 4s.‡

117. **SUBTERRANEOUS SURVEYING**, an Elementary and Practical Treatise on. By THOMAS FENWICK. Also the Method of Conducting Subterraneous Surveys without the Use of the Magnetic Needle, and other Modern Improvements. By THOMAS BAKER, C.E. Illustrated. 2s. 6d.‡

118. **CIVIL ENGINEERING IN NORTH AMERICA**, a Sketch of. By DAVID STEVENSON, F.R.S.E., &c. Plates and Diagrams. 3s.

197. **ROADS AND STREETS (THE CONSTRUCTION OF)**, in two Parts: I. THE ART OF CONSTRUCTING COMMON ROADS, by HENRY LAW, C.E., revised by D. K. CLARK, C.E.; II. RECENT PRACTICE, including pavements of Stone, Wood, and Asphalte, by D. K. CLARK. 4s. 6d.‡

203. **SANITARY WORK IN THE SMALLER TOWNS AND IN VILLAGES.** Comprising:—1. Some of the more Common Forms of Nuisance and their Remedies; 2. Drainage; 3. Water Supply. By CHARLES SLAGG, A.I.C.E. 2s. 6d.‡

212. **THE CONSTRUCTION OF GAS-WORKS**, and the Manufacture and Distribution of Coal Gas. Originally written by SAMUEL HUGHES, C.E. Sixth Edition, re-written and much Enlarged by WILLIAM RICHARDS, C.E. With 72 Illustrations. 4s. 6d.‡

213. **PIONEER ENGINEERING.** A Treatise on the Engineering Operations connected with the Settlement of Waste Lands in New Countries. By EDWARD DOBSON, Assoc. Inst. C.E. 4s. 6d.‡

☞ *The ‡ indicates that these vols. may be had strongly bound at 6d. extra.*

LONDON : CROSBY LOCKWOOD AND CO.,

MECHANICAL ENGINEERING, ETC.

33. *CRANES*, the Construction of, and other Machinery for Raising Heavy Bodies for the Erection of Buildings, and for Hoisting Goods. By JOSEPH GLYNN, F.R.S., &c. Illustrated. 1s. 6d.
34. *THE STEAM ENGINE*. By Dr. LARDNER. Illustrated. 1s. 6d.
59. *STEAM BOILERS*: their Construction and Management. By R. ARMSTRONG, C.E. Illustrated. 1s. 6d.
67. *CLOCKS, WATCHES, AND BELLS*, a Rudimentary Treatise on. By Sir EDMUND BECKETT (late EDMUND BECKETT DENISON), LL.D., Q.C. A New, Revised, and considerably Enlarged Edition (the 6th), with very numerous Illustrations. 4s. 6d. cloth limp; 5s. 6d. cloth boards, gilt.
82. *THE POWER OF WATER*, as applied to drive Flour Mills, and to give motion to Turbines and other Hydrostatic Engines. By JOSEPH GLYNN, F.R.S., &c. New Edition, Illustrated. 2s.‡
98. *PRACTICAL MECHANISM*, the Elements of; and Machine Tools. By T. BAKER, C.E. With Additions by J. NASMYTH, C.E. 2s. 6d.‡
139. *THE STEAM ENGINE*, a Treatise on the Mathematical Theory of, with Rules and Examples for Practical Men. By T. BAKER, C.E. 1s. 6d.
162. *THE BRASS FOUNDER'S MANUAL;* Instructions for Modelling, Pattern-Making, Moulding, Turning, Filing, Burnishing, Bronzing, &c. With copious Receipts, &c. By WALTER GRAHAM. 2s.‡
164. *MODERN WORKSHOP PRACTICE*, as applied to Marine, Land, and Locomotive Engines, Floating Docks, Dredging Machines, Bridges, Cranes, Ship-building, &c., &c. By J. G. WINTON. Illustrated. 3s.‡
165. *IRON AND HEAT*, exhibiting the Principles concerned in the Construction of Iron Beams, Pillars, and Bridge Girders, and the Action of Heat in the Smelting Furnace. By J. ARMOUR, C.E. 2s. 6d.‡
166. *POWER IN MOTION:* Horse-Power, Toothed-Wheel Gearing, Long and Short Driving Bands, and Angular Forces. By J. ARMOUR, 2s. 6d.‡
167. *IRON BRIDGES, GIRDERS, ROOFS, AND OTHER WORKS*. By FRANCIS CAMPIN, C.E. 2s. 6d.‡
171. *THE WORKMAN'S MANUAL OF ENGINEERING DRAWING*. By JOHN MAXTON, Engineer. Fourth Edition. Illustrated with 7 Plates and nearly 350 Woodcuts. 3s. 6d.‡
190. *STEAM AND THE STEAM ENGINE*, Stationary and Portable. Being an extension of Mr. John Sewell's "Treatise on Steam." By D. K. CLARK, M.I.C.E. Second Edition, revised. 3s. 6d.‡
200. *FUEL*, its Combustion and Economy. By C. W. WILLIAMS, A.I.C.E. With extensive additions on Recent Practice in the Combustion and Economy of Fuel—Coal, Coke, Wood, Peat, Petroleum, &c.—by D. K. CLARK, M.I.C.E. 2nd Edition. 3s. 6d.‡
202. *LOCOMOTIVE ENGINES*. By G. D. DEMPSEY, C.E.; with large additions by D. KINNEAR CLARK, M.I.C.E. 3s.‡
211. *THE BOILERMAKER'S ASSISTANT* in Drawing, Templating, and Calculating Boiler and Tank Work. By JOHN COURTNEY, Practical Boiler Maker. Edited by D. K. CLARK, C.E. 100 Illustrations. 2s.
216. *MATERIALS AND CONSTRUCTION;* A Theoretical and Practical Treatise on the Strains, Designing, and Erection of Works of Construction. By FRANCIS CAMPIN, C.E. 3s.‡ [*Just published.*
217. *SEWING MACHINERY:* a Manual of the Sewing Machine; comprising its Construction, History, &c., with full Technical Directions for Adjusting, &c. By J. W. URQUHART, C.E. 2s.‡
223. *MECHANICAL ENGINEERING*. Comprising Metallurgy, Moulding, Casting, Forging, Tools, Workshop Machinery, Manufacture of the Steam Engine, &c. By FRANCIS CAMPIN, C.E. 2s. 6d.‡
224. *COACH BUILDING*, A Practical Treatise, Historical and Descriptive, containing full information of the various Trades and Processes involved, with Hints on the proper Keeping of Carriages. By J. W. BURGESS. 2s. 6d.‡
PRACTICAL ORGAN BUILDING. By W. E. DICKSON, M.A., Precentor of Ely Cathedral. Illustrated. 2s. 6d.‡ [*Just published.*

☞ *The ‡ indicates that these vols. may be had strongly bound at 6d. extra.*

SHIPBUILDING, NAVIGATION, MARINE ENGINEERING, ETC.

51. *NAVAL ARCHITECTURE*, the Rudiments of; or an Exposition of the Elementary Principles of the Science, and their Practical Application to Naval Construction. Compiled for the Use of Beginners. By JAMES PEAKE, School of Naval Architecture, H.M. Dockyard, Portsmouth. Fourth Edition, corrected, with Plates and Diagrams. 3s. 6d.‡

53*. *SHIPS FOR OCEAN AND RIVER SERVICE*, Elementary and Practical Principles of the Construction of. By HAKON A. SOMMERFELDT, Surveyor of the Royal Norwegian Navy. With an Appendix. 1s. 6d.

53**. *AN ATLAS OF ENGRAVINGS* to Illustrate the above. Twelve large folding plates. Royal 4to, cloth. 7s. 6d.

54. *MASTING, MAST-MAKING, AND RIGGING OF SHIPS*, Rudimentary Treatise on. Also Tables of Spars, Rigging, Blocks; Chain, Wire, and Hemp Ropes, &c., relative to every class of vessels. With an Appendix of Dimensions of Masts and Yards of the Royal Navy. By ROBERT KIPPING, N.A. Fourteenth Edition. Illustrated. 2s.‡

54*. *IRON SHIP-BUILDING*. With Practical Examples and Details for the Use of Ship Owners and Ship Builders. By JOHN GRANTHAM, Consulting Engineer and Naval Architect. 5th Edition, with Additions. 4s.

54**. *AN ATLAS OF FORTY PLATES* to Illustrate the above. Fifth Edition. Including the latest Examples, such as H.M. Steam Frigates "Warrior," "Hercules," "Bellerophon;" H.M. Troop Ship "Serapis," Iron Floating Dock, &c., &c. 4to, boards. 38s.

55. *THE SAILOR'S SEA BOOK:* a Rudimentary Treatise on Navigation. Part I. How to Keep the Log and Work it off. Part II. On Finding the Latitude and Longitude. By JAMES GREENWOOD, B.A. To which are added, the Deviation and Error of the Compass; Great Circle Sailing; the International (Commercial) Code of Signals; the Rule of the Road at Sea; Rocket and Mortar Apparatus for Saving Life; the Law of Storms; and a Brief Dictionary of Sea Terms. With numerous Woodcuts and Coloured Plates of Flags. New, thoroughly revised and much enlarged edition. By W. H. ROSSER. 2s. 6d.‡

80. *MARINE ENGINES, AND STEAM VESSELS*, a Treatise on. Together with Practical Remarks on the Screw and Propelling Power, as used in the Royal and Merchant Navy. By ROBERT MURRAY, C.E., Engineer-Surveyor to the Board of Trade. With a Glossary of Technical Terms, and their Equivalents in French, German, and Spanish. Seventh Edition, revised and enlarged. Illustrated. 3s.‡

83*bis*. *THE FORMS OF SHIPS AND BOATS:* Hints, Experimentally Derived, on some of the Principles regulating Ship-building. By W. BLAND. Seventh Edition, revised, with numerous Illustrations and Models. 1s.6d.

99. *NAVIGATION AND NAUTICAL ASTRONOMY*, in Theory and Practice. With Attempts to facilitate the Finding of the Time and the Longitude at Sea. By J. R. YOUNG, formerly Professor of Mathematics in Belfast College. Illustrated. 2s. 6d.

100*. *TABLES* intended to facilitate the Operations of Navigation and Nautical Astronomy, as an Accompaniment to the above Book. By J. R. YOUNG. 1s. 6d.

106. *SHIPS' ANCHORS*, a Treatise on. By G. COTSELL, N.A. 1s. 6d.

149. *SAILS AND SAIL-MAKING*, an Elementary Treatise on. With Draughting, and the Centre of Effort of the Sails. Also, Weights and Sizes of Ropes; Masting, Rigging, and Sails of Steam Vessels, &c., &c. Eleventh Edition, enlarged, with an Appendix. By ROBERT KIPPING, N.A., Sailmaker, Quayside, Newcastle. Illustrated. 2s. 6d.‡

155. *THE ENGINEER'S GUIDE TO THE ROYAL AND MERCANTILE NAVIES*. By a PRACTICAL ENGINEER. Revised by D. F. M'CARTHY, late of the Ordnance Survey Office, Southampton. 3s.

55 & 204. *PRACTICAL NAVIGATION*. Consisting of The Sailor's Sea-Book. By JAMES GREENWOOD and W. H. ROSSER. Together with the requisite Mathematical and Nautical Tables for the Working of the Problems. By HENRY LAW, C.E., and J. R. YOUNG, formerly Professor of Mathematics in Belfast College. Illustrated with numerous Wood Engravings and Coloured Plates. 7s. Strongly half-bound in leather.

☞ *The ‡ indicates that these vols. may be had strongly bound at 6d. extra.*

PHYSICAL SCIENCE, NATURAL PHILOSOPHY, ETC.

1. *CHEMISTRY*, for the Use of Beginners. By Professor GEORGE FOWNES, F.R.S. With an Appendix on the Application of Chemistry to Agriculture. 1s.
2. *NATURAL PHILOSOPHY*, Introduction to the Study of; for the Use of Beginners. By C. TOMLINSON, Lecturer on Natural Science in King's College School, London. Woodcuts. 1s. 6d.
4. *MINERALOGY*, Rudiments of; a concise View of the Properties of Minerals. By A. RAMSAY, Jun. Woodcuts and Steel Plates. 3s.‡
6. *MECHANICS*, Rudimentary Treatise on; being a concise Exposition of the General Principles of Mechanical Science, and their Applications. By CHARLES TOMLINSON. Illustrated. 1s. 6d.
7. *ELECTRICITY;* showing the General Principles of Electrical Science, and the purposes to which it has been applied. By Sir W. SNOW HARRIS, F.R.S., &c. With Additions by R. SABINE, C.E., F.S.A. 1s. 6d.
7*. *GALVANISM*, Rudimentary Treatise on, and the General Principles of Animal and Voltaic Electricity. By Sir W. SNOW HARRIS. New Edition, with considerable Additions by ROBERT SABINE, C.E., F.S.A. 1s. 6d.
8. *MAGNETISM;* being a concise Exposition of the General Principles of Magnetical Science, and the Purposes to which it has been applied. By Sir W. SNOW HARRIS. New Edition, revised and enlarged by H. M. NOAD, Ph.D., Vice-President of the Chemical Society, Author of "A Manual of Electricity," &c., &c. With 165 Woodcuts. 3s. 6d.‡
11. *THE ELECTRIC TELEGRAPH;* its History and Progress; with Descriptions of some of the Apparatus. By R. SABINE, C.E., F.S.A. 3s.
12. *PNEUMATICS*, for the Use of Beginners. By CHARLES TOMLINSON. Illustrated. 1s. 6d.
72. *MANUAL OF THE MOLLUSCA;* a Treatise on Recent and Fossil Shells. By Dr. S. P. WOODWARD, A.L.S. Fourth Edition. With Appendix by RALPH TATE, A.L.S., F.G.S. With numerous Plates and 300 Woodcuts. 6s. 6d. Cloth boards, 7s. 6d.
79**. *PHOTOGRAPHY*, Popular Treatise on; with a Description of the Stereoscope, &c. Translated from the French of D. VAN MONCKHOVEN, by W. H. THORNTHWAITE, Ph.D. Woodcuts. 1s. 6d.
96. *ASTRONOMY.* By the Rev. R. MAIN, M.A., F.R.S., &c. New Edition, with an Appendix on "Spectrum Analysis." Woodcuts. 1s. 6d.
97. *STATICS AND DYNAMICS*, the Principles and Practice of; embracing also a clear development of Hydrostatics, Hydrodynamics, and Central Forces. By T. BAKER, C.E. 1s. 6d.
138. *TELEGRAPH*, Handbook of the; a Manual of Telegraphy, Telegraph Clerks' Remembrancer, and Guide to Candidates for Employment in the Telegraph Service. By R. BOND. Fourth Edition, revised and enlarged: to which is appended, QUESTIONS on MAGNETISM, ELECTRICITY, and PRACTICAL TELEGRAPHY, for the Use of Students, by W. McGREGOR, First Assistant Supnt., Indian Gov. Telegraphs. 3s.‡
143. *EXPERIMENTAL ESSAYS.* By CHARLES TOMLINSON. I. On the Motions of Camphor on Water. II. On the Motion of Camphor towards the Light. III. History of the Modern Theory of Dew. Woodcuts. 1s.
173. *PHYSICAL GEOLOGY*, partly based on Major-General PORTLOCK's "Rudiments of Geology." By RALPH TATE, A.L.S., &c. Woodcuts. 2s.
174. *HISTORICAL GEOLOGY*, partly based on Major-General PORTLOCK's "Rudiments." By RALPH TATE, A.L.S., &c. Woodcuts. 2s. 6d.
173 & 174. *RUDIMENTARY TREATISE ON GEOLOGY*, Physical and Historical. Partly based on Major-General PORTLOCK's "Rudiments of Geology." By RALPH TATE, A.L.S., F.G.S., &c. In One Volume. 4s. 6d.‡
183 & 184. *ANIMAL PHYSICS*, Handbook of. By Dr. LARDNER, D.C.L., formerly Professor of Natural Philosophy and Astronomy in University College, Lond. With 520 Illustrations. In One Vol. 7s. 6d., cloth boards.

*** *Sold also in Two Parts, as follows:*—

183. ANIMAL PHYSICS. By Dr. LARDNER. Part I., Chapters I.—VII. 4s.
184. ANIMAL PHYSICS. By Dr. LARDNER. Part II., Chapters VIII.—XVIII. 3s.

☞ *The ‡ indicates that these vols. may be had strongly bound at 6d. extra.*

7, STATIONERS' HALL COURT, LUDGATE HILL, E.C.

MINING, METALLURGY, ETC.

117. *SUBTERRANEOUS SURVEYING*, Elementary and Practical Treatise on, with and without the Magnetic Needle. By THOMAS FENWICK, Surveyor of Mines, and THOMAS BAKER, C.E. Illustrated. 2s. 6d.‡
133. *METALLURGY OF COPPER;* an Introduction to the Methods of Seeking, Mining, and Assaying Copper, and Manufacturing its Alloys. By ROBERT H. LAMBORN, Ph.D. Woodcuts. 2s. 6d.‡
134. *METALLURGY OF SILVER AND LEAD.* A Description of the Ores; their Assay and Treatment, and valuable Constituents. By Dr. R. H. LAMBORN. Woodcuts. 2s. 6d.‡
135. *ELECTRO-METALLURGY;* Practically Treated. By ALEXANDER WATT, F.R.S.S.A. 7th Edition, revised, with important additions, including the Electro-Deposition of Nickel, &c. Woodcuts. 3s.‡
172. *MINING TOOLS*, Manual of. For the Use of Mine Managers, Agents, Students, &c. By WILLIAM MORGANS. 2s. 6d.‡
172*. *MINING TOOLS, ATLAS* of Engravings to Illustrate the above, containing 235 Illustrations, drawn to Scale. 4to. 4s. 6d.; cloth boards, 6s.
176. *METALLURGY OF IRON.* Containing History of Iron Manufacture, Methods of Assay, and Analyses of Iron Ores, Processes of Manufacture of Iron and Steel, &c. By H. BAUERMAN, F.G.S. 4th Edition. 4s. 6d.‡
180. *COAL AND COAL MINING*, A Rudimentary Treatise on. By WARINGTON W. SMYTH, M.A., F.R.S. Fifth Edition, revised and enlarged. With numerous Illustrations. 3s. 6d.‡ [*Just published.*
195. *THE MINERAL SURVEYOR AND VALUER'S COMPLETE GUIDE,* with new Traverse Tables, and Descriptions of Improved Instruments; also the Correct Principles of Laying out and Valuing Mineral Properties. By WILLIAM LINTERN, Mining and Civil Engineer. 3s. 6d.‡
214. *SLATE AND SLATE QUARRYING*, Scientific, Practical, and Commercial. By D. C. DAVIES, F.G.S., Mining Engineer, &c. With numerous Illustrations and Folding Plates. 3s.‡
215. *THE GOLDSMITH'S HANDBOOK*, containing full Instructions for the Alloying and Working of Gold. By GEORGE E. GEE, Goldsmith and Silversmith. Second Edition, considerably enlarged. 3s.‡[*Just published.*
225. *THE SILVERSMITH'S HANDBOOK*, containing full Instructions for the Alloying and Working of Silver. By GEORGE E. GEE. 3s.‡
220. *MAGNETIC SURVEYING, AND ANGULAR SURVEYING*, with Records of the Peculiarities of Needle Disturbances. Compiled from the Results of carefully made Experiments. By WILLIAM LINTERN, Mining and Civil Engineer and Surveyor. 2s. [*Just published.*

FINE ARTS.

20. *PERSPECTIVE FOR BEGINNERS.* Adapted to Young Students and Amateurs in Architecture, Painting. &c. By GEORGE PYNE. 2°.
40 *GLASS STAINING, AND THE ART OF PAINTING ON*
& 41. *GLASS.* From the German of Dr. GESSERT and EMANUEL OTTO FROMBERG. With an Appendix on THE ART OF ENAMELLING. 2s. 6d.
69. *MUSIC*, A Rudimentary and Practical Treatise on. With numerous Examples. By CHARLES CHILD SPENCER. 2s. 6d.
71. *PIANOFORTE*, The Art of Playing the. With numerous Exercises & Lessons from the Best Masters. By CHARLES CHILD SPENCER. 1s.6d.
69-71. *MUSIC AND THE PIANOFORTE.* In one volume. Half bound, 5s.
181. *PAINTING POPULARLY EXPLAINED*, including Fresco, Oil, Mosaic, Water Colour, Water-Glass, Tempera, Encaustic, Miniature, Painting on Ivory, Vellum, Pottery, Enamel, Glass, &c. With Historical Sketches of the Progress of the Art by THOMAS JOHN GULLICK, assisted by JOHN TIMBS, F.S.A. Fourth Edition, revised and enlarged. 5s.‡
186. *A GRAMMAR OF COLOURING*, applied to Decorative Painting and the Arts. By GEORGE FIELD. New Edition, enlarged and adapted to the Use of the Ornamental Painter and Designer. By ELLIS A. DAVIDSON. With two new Coloured Diagrams, &c. 3s.‡

☞ *The ‡ indicates that these vols. may be had strongly bound at 6d. extra.*

LONDON: CROSBY LOCKWOOD AND CO.,

AGRICULTURE, GARDENING, ETC.

66. *CLAY LANDS & LOAMY SOILS.* By Prof. DONALDSON. 1s.
131. *MILLER'S, MERCHANT'S, AND FARMER'S READY* RECKONER, for ascertaining at sight the value of any quantity of Corn, from One Bushel to One Hundred Quarters, at any given price, from £1 to £5 per Qr. With approximate values of Millstones, Millwork, &c. 1s.
140. *SOILS, MANURES, AND CROPS.* (Vol. 1. OUTLINES OF MODERN FARMING.) By R. SCOTT BURN. Woodcuts. 2s.
141. *FARMING & FARMING ECONOMY,* Notes, Historical and Practical, on. (Vol. 2. OUTLINES OF MODERN FARMING.) By R. SCOTT BURN. 3s.
142. *STOCK; CATTLE, SHEEP, AND HORSES.* (Vol. 3. OUTLINES OF MODERN FARMING.) By R. SCOTT BURN. Woodcuts. 2s. 6d.
145. *DAIRY, PIGS, AND POULTRY,* Management of the. By R. SCOTT BURN. With Notes on the Diseases of Stock. (Vol. 4. OUTLINES OF MODERN FARMING.) Woodcuts. 2s.
146. *UTILIZATION OF SEWAGE, IRRIGATION, AND RECLAMATION OF WASTE LAND.* (Vol. 5. OUTLINES OF MODERN FARMING.) By R. SCOTT BURN. Woodcuts. 2s. 6d.
*** Nos. 140-1-2-5-6, in One Vol., handsomely half-bound, entitled "OUTLINES OF MODERN FARMING." By ROBERT SCOTT BURN. Price 12s.
177. *FRUIT TREES,* The Scientific and Profitable Culture of. From the French of DU BREUIL. Revised by GEO. GLENNY. 187 Woodcuts. 3s. 6d.‡
198. *SHEEP; THE HISTORY, STRUCTURE, ECONOMY, AND DISEASES OF.* By W. C. SPOONER, M.R.V.C., &c. Fourth Edition, enlarged, including Specimens of New and Improved Breeds. 3s. 6d.‡
201. *KITCHEN GARDENING MADE EASY.* Showing how to prepare and lay out the ground, the best means of cultivating every known Vegetable and Herb, with full cultural directions, &c. By GEORGE M. F. GLENNY. 1s. 6d.‡
207. *OUTLINES OF FARM MANAGEMENT,* and the Organization *of Farm Labour:* Treating of the General Work of the Farm; Field and Live Stock; Details of Contract Work; Specialities of Labour, &c., &c. By ROBERT SCOTT BURN. 2s. 6d.‡
208. *OUTLINES OF LANDED ESTATES MANAGEMENT:* Treating of the Varieties of Lands, Methods of Farming, Farm Buildings, Irrigation, Drainage, &c. By R. SCOTT BURN. 2s. 6d.‡
*** *Nos.* 207 & 208 *in One Vol., handsomely half-bound, entitled* "OUTLINES OF LANDED ESTATES AND FARM MANAGEMENT." By R. SCOTT BURN. *Price* 6s.
209. *THE TREE PLANTER AND PLANT PROPAGATOR.* A Practical Manual on the Propagation of Forest Trees, Fruit Trees, Flowering Shrubs, Flowering Plants, Pot-Herbs, &c. By SAMUEL WOOD. Illustrated. 2s.‡
210. *THE TREE PRUNER.* A Practical Manual on the Pruning of Fruit Trees, including also their Training and Renovation; also the Pruning of Shrubs, Climbers, and Flowering Plants. By SAMUEL WOOD. 2s.‡
*** *Nos.* 209 & 210 *in One Vol., handsomely half-bound, entitled* "THE TREE PLANTER, PROPAGATOR AND PRUNER." By SAMUEL WOOD. *Price* 5s.
219. *THE HAY AND STRAW MEASURER:* Being New Tables for the Use of Auctioneers, Valuers, Farmers, Hay and Straw Dealers, &c., forming a complete Calculator and Ready-Reckoner, especially adapted to persons connected with Agriculture. Fourth Edition. By JOHN STEELE. 2s.
222. *SUBURBAN FARMING.* The Laying-out and Cultivation of Farms, adapted to the Produce of Milk, Butter, and Cheese, Eggs, Poultry, and Pigs. By Prof. JOHN DONALDSON and R. SCOTT BURN. 3s. 6d.‡
231. *THE ART OF GRAFTING AND BUDDING.* By CHARLES BALTET. With Illustrations. 2s. 6d.‡ *[Just published.*
232. *COTTAGE GARDENING;* or, Flowers, Fruits, and Vegetables for Small Gardens. By E. HOBDAY. 1s. 6d. *[Just published.*
233. *GARDEN RECEIPTS.* Edited by CHARLES W. QUIN. 1s. 6d. *[Just published.*

☞ *The* ‡ *indicates that these vols. may be had strongly bound at* 6d. *extra.*

7, STATIONERS' HALL COURT, LUDGATE HILL, E.C.

ARITHMETIC, GEOMETRY, MATHEMATICS, ETC.

32. *MATHEMATICAL INSTRUMENTS*, a Treatise on; in which their Construction and the Methods of Testing, Adjusting, and Using them are concisely Explained. By J. F. HEATHER, M.A., of the Royal Military Academy, Woolwich. Original Edition, in 1 vol., Illustrated. 1s. 6d.

⁎ In ordering the above, be careful to say, "*Original Edition*" (*No.* 32), *to distinguish it from the Enlarged Edition in* 3 *vols.* (*Nos.* 168-9-70.)

60. *LAND AND ENGINEERING SURVEYING*, a Treatise on; with all the Modern Improvements. Arranged for the Use of Schools and Private Students; also for Practical Land Surveyors and Engineers. By T. BAKER, C.E. New Edition, revised by EDWARD NUGENT, C.E. Illustrated with Plates and Diagrams. 2s.‡

61*. *READY RECKONER FOR THE ADMEASUREMENT OF LAND.* By ABRAHAM ARMAN, Schoolmaster, Thurleigh, Beds. To which is added a Table, showing the Price of Work, from 2s. 6d. to £1 per acre, and Tables for the Valuation of Land, from 1s. to £1,000 per acre, and from one pole to two thousand acres in extent, &c., &c. 1s. 6d.

76. *DESCRIPTIVE GEOMETRY*, an Elementary Treatise on; with a Theory of Shadows and of Perspective, extracted from the French of G. MONGE. To which is added, a description of the Principles and Practice of Isometrical Projection; the whole being intended as an introduction to the Application of Descriptive Geometry to various branches of the Arts. By J. F. HEATHER, M.A. Illustrated with 14 Plates. 2s.

178. *PRACTICAL PLANE GEOMETRY:* giving the Simplest Modes of Constructing Figures contained in one Plane and Geometrical Construction of the Ground. By J. F. HEATHER, M.A. With 215 Woodcuts. 2s.

179. *PROJECTION :* Orthographic, Topographic, and Perspective: giving the various Modes of Delineating Solid Forms by Constructions on a Single Plane Surface. By J. F. HEATHER, M.A. [*In preparation.*

⁎ The above three volumes will form a COMPLETE ELEMENTARY COURSE OF MATHEMATICAL DRAWING.

83. *COMMERCIAL BOOK-KEEPING.* With Commercial Phrases and Forms in English, French, Italian, and German. By JAMES HADDON, M.A., Arithmetical Master of King's College School, London. 1s. 6d.

84. *ARITHMETIC*, a Rudimentary Treatise on: with full Explanations of its Theoretical Principles, and numerous Examples for Practice. For the Use of Schools and for Self-Instruction. By J. R. YOUNG, late Professor of Mathematics in Belfast College. New Edition, with Index. 1s. 6d.

84*. A KEY to the above, containing Solutions in full to the Exercises, together with Comments, Explanations, and Improved Processes, for the Use of Teachers and Unassisted Learners. By J. R. YOUNG. 1s. 6d.

85. *EQUATIONAL ARITHMETIC*, applied to Questions of Interest, 85*. Annuities, Life Assurance, and General Commerce; with various Tables by which all Calculations may be greatly facilitated. By W. HIPSLEY. 2s.

86. *ALGEBRA*, the Elements of. By JAMES HADDON, M.A., Second Mathematical Master of King's College School. With Appendix, containing miscellaneous Investigations, and a Collection of Problems in various parts of Algebra. 2s.

86*. A KEY AND COMPANION to the above Book, forming an extensive repository of Solved Examples and Problems in Illustration of the various Expedients necessary in Algebraical Operations. Especially adapted for Self-Instruction. By J. R. YOUNG. 1s. 6d.

88. *EUCLID*, THE ELEMENTS OF : with many additional Propositions
89. and Explanatory Notes: to which is prefixed, an Introductory Essay on Logic. By HENRY LAW, C.E. 2s. 6d.‡

⁎ *Sold also separately, viz. :—*

88. EUCLID, The First Three Books. By HENRY LAW, C.E. 1s. 6d.
89. EUCLID, Books 4, 5, 6, 11, 12. By HENRY LAW, C.E. 1s. 6d.

☞ *The ‡ indicates that these vols. may be had strongly bound at* 6d. *extra.*

LONDON: CROSBY LOCKWOOD AND CO.,

Arithmetic, Geometry, Mathematics, etc., *continued*.

90. *ANALYTICAL GEOMETRY AND CONIC SECTIONS*, a Rudimentary Treatise on. By JAMES HANN, late Mathematical Master of King's College School, London. A New Edition, re-written and enlarged by J. R. YOUNG, formerly Professor of Mathematics at Belfast College. 2s.‡

91. *PLANE TRIGONOMETRY*, the Elements of. By JAMES HANN, formerly Mathematical Master of King's College, London. 1s. 6d.

92. *SPHERICAL TRIGONOMETRY*, the Elements of. By JAMES HANN. Revised by CHARLES H. DOWLING, C.E. 1s.
*** Or with "*The Elements of Plane Trigonometry*," in *One Volume*, 2s. 6d.

93. *MENSURATION AND MEASURING*, for Students and Practical Use. With the Mensuration and Levelling of Land for the Purposes of Modern Engineering. By T. BAKER, C.E. New Edition, with Corrections and Additions by E. NUGENT, C.E. Illustrated. 1s. 6d.

102. *INTEGRAL CALCULUS*, Rudimentary Treatise on the. By HOMERSHAM COX, B.A. Illustrated. 1s.

103. *INTEGRAL CALCULUS*, Examples on the. By JAMES HANN, late of King's College, London. Illustrated. 1s.

101. *DIFFERENTIAL CALCULUS*, Elements of the. By W. S. B. WOOLHOUSE, F.R.A.S., &c. 1s. 6d.

105. *MNEMONICAL LESSONS.* — GEOMETRY, ALGEBRA, AND TRIGONOMETRY, in Easy Mnemonical Lessons. By the Rev. THOMAS PENYNGTON KIRKMAN, M.A. 1s. 6d.

136. *ARITHMETIC*, Rudimentary, for the Use of Schools and Self-Instruction. By JAMES HADDON, M.A. Revised by ABRAHAM ARMAN. 1s. 6d.

137. A KEY TO HADDON'S RUDIMENTARY ARITHMETIC. By A. ARMAN. 1s. 6d.

168. *DRAWING AND MEASURING INSTRUMENTS.* Including—I. Instruments employed in Geometrical and Mechanical Drawing, and in the Construction, Copying, and Measurement of Maps and Plans. II. Instruments used for the purposes of Accurate Measurement, and for Arithmetical Computations. By J. F. HEATHER, M.A., late of the Royal Military Academy, Woolwich, Author of "Descriptive Geometry," &c., &c. Illustrated. 1s. 6d.

169. *OPTICAL INSTRUMENTS.* Including (more especially) Telescopes, Microscopes, and Apparatus for producing copies of Maps and Plans by Photography. By J. F. HEATHER, M.A. Illustrated. 1s. 6d.

170. *SURVEYING AND ASTRONOMICAL INSTRUMENTS.* Including—I. Instruments Used for Determining the Geometrical Features of a portion of Ground. II. Instruments Employed in Astronomical Observations. By J. F. HEATHER, M.A. Illustrated. 1s. 6d.

*** *The above three volumes form an enlargement of the Author's original work,* "*Mathematical Instruments: their Construction, Adjustment, Testing, and Use,*" *the Thirteenth Edition of which is on sale, price* 1s. 6d. (*See No.* 32 *in the Series.*)

168.⎫
169.⎬ *MATHEMATICAL INSTRUMENTS.* By J. F. HEATHER, M.A. Enlarged Edition, for the most part entirely re-written. The 3 Parts as
170.⎭ above, in One thick Volume. With numerous Illustrations. 4s. 6d.‡

158. *THE SLIDE RULE, AND HOW TO USE IT;* containing full, easy, and simple Instructions to perform all Business Calculations with unexampled rapidity and accuracy. By CHARLES HOARE, C.E. With a Slide Rule in tuck of cover. 2s. 6d.‡

185. *THE COMPLETE MEASURER*; setting forth the Measurement of Boards, Glass, &c., &c.; Unequal-sided, Square-sided, Octagonal-sided, Round Timber and Stone, and Standing Timber. With a Table showing the solidity of hewn or eight-sided timber, or of any octagonal-sided column. Compiled for Timber-growers, Merchants, and Surveyors, Stonemasons, Architects, and others. By RICHARD HORTON. Fourth Edition, with valuable additions. 4s.; strongly bound in leather, 5s.

196. *THEORY OF COMPOUND INTEREST AND ANNUITIES;* with Tables of Logarithms for the more Difficult Computations of Interest, Discount, Annuities, &c. By FÉDOR THOMAN. 4s.‡

☞ *The* ‡ *indicates that these vols. may be had strongly bound at* 6d. *extra.*

7, STATIONERS' HALL COURT, LUDGATE HILL, E.C.

WEALE'S RUDIMENTARY SERIES.

Arithmetic, Geometry, Mathematics, etc., *continued*.

199. *INTUITIVE CALCULATIONS;* or, Easy and Compendious Methods of Performing the various Arithmetical Operations required in Commercial and Business Transactions; together with Full Explanations of Decimals and Duodecimals, several Useful Tables, &c. By DANIEL O'GORMAN. Twenty-fifth Edition, corrected and enlarged by J. R. YOUNG, formerly Professor of Mathematics in Belfast College. 3s.‡

204. *MATHEMATICAL TABLES,* for Trigonometrical, Astronomical, and Nautical Calculations; to which is prefixed a Treatise on Logarithms. By HENRY LAW, C.E. Together with a Series of Tables for Navigation and Nautical Astronomy. By J. R. YOUNG, formerly Professor of Mathematics in Belfast College. New Edition. 3s. 6d.‡

221. *MEASURES, WEIGHTS, AND MONEYS OF ALL NATIONS,* and an Analysis of the Christian, Hebrew, and Mahometan Calendars. By W. S. B. WOOLHOUSE, F.R.A.S., F.S.S. Sixth Edition, carefully revised and enlarged. 2s.‡

227. *MATHEMATICS AS APPLIED TO THE CONSTRUCTIVE ARTS.* Illustrating the various processes of Mathematical Investigation, by means of Arithmetical and Simple Algebraical Equations and Practical Examples; also the Methods of Analysing Principles and Deducing Rules and Formulæ, applicable to the Requirements of Practice. By FRANCIS CAMPIN, C.E., Author of "Materials and Construction," &c. Second Edition, revised and enlarged by the Author. 3s.‡ [*Just published.*

MISCELLANEOUS VOLUMES.

36. *A DICTIONARY OF TERMS used in ARCHITECTURE, BUILDING, ENGINEERING, MINING, METALLURGY, ARCHÆOLOGY, the FINE ARTS, &c.* By JOHN WEALE. Fifth Edition. Revised by ROBERT HUNT, F.R.S., Keeper of Mining Records. Numerous Illustrations. 5s. cloth limp; 6s. cloth boards.

50. *THE LAW OF CONTRACTS FOR WORKS AND SERVICES.* By DAVID GIBBONS. Third Edition, enlarged. 3s.‡

112. *MANUAL OF DOMESTIC MEDICINE.* By R. GOODING, B.A., M.D. Intended as a Family Guide in all Cases of Accident and Emergency. 2s.‡

112*. *MANAGEMENT OF HEALTH.* A Manual of Home and Personal Hygiene. By the Rev. JAMES BAIRD, B.A. 1s.

150. *LOGIC,* Pure and Applied. By S. H. EMMENS. 1s. 6d.

153. *SELECTIONS FROM LOCKE'S ESSAYS ON THE HUMAN UNDERSTANDING.* With Notes by S. H. EMMENS. 2s.

154. *GENERAL HINTS TO EMIGRANTS.* Containing Notices of the various Fields for Emigration. With Hints on Preparation for Emigrating, Outfits, &c., &c. With Directions and Recipes useful to the Emigrant. With a Map of the World. 2s.

157. *THE EMIGRANT'S GUIDE TO NATAL.* By ROBERT JAMES MANN, F.R.A.S., F.M.S. Second Edition, carefully corrected to the present Date. Map. 2s.

193. *HANDBOOK OF FIELD FORTIFICATION,* intended for the Guidance of Officers Preparing for Promotion, and especially adapted to the requirements of Beginners. By Major W. W. KNOLLYS, F.R.G.S., 93rd Sutherland Highlanders, &c. With 163 Woodcuts. 3s.‡

194. *THE HOUSE MANAGER:* Being a Guide to Housekeeping. Practical Cookery, Pickling and Preserving, Household Work, Dairy Management, the Table and Dessert, Cellarage of Wines, Home-brewing and Wine-making, the Boudoir and Dressing-room, Travelling, Stable Economy, Gardening Operations, &c. By AN OLD HOUSEKEEPER. 3s. 6d.‡

194. *HOUSE BOOK (The).* Comprising :—I. THE HOUSE MANAGER.
112. By an OLD HOUSEKEEPER. II. DOMESTIC MEDICINE. By RALPH GOODING,
& M.D. III. MANAGEMENT OF HEALTH. By JAMES BAIRD. In One Vol.,
112*. strongly half-bound. 6s.

The ‡ indicates that these vols. may be had strongly bound at 6d. extra.

LONDON : CROSBY LOCKWOOD AND CO.,

EDUCATIONAL AND CLASSICAL SERIES.

HISTORY.

1. **England, Outlines of the History of;** more especially with reference to the Origin and Progress of the English Constitution. By WILLIAM DOUGLAS HAMILTON, F.S.A., of Her Majesty's Public Record Office. 4th Edition, revised. 5s.; cloth boards, 6s.
5. **Greece, Outlines of the History of;** in connection with the Rise of the Arts and Civilization in Europe. By W. DOUGLAS HAMILTON, of University College, London, and EDWARD LEVIEN, M.A., of Balliol College, Oxford. 2s. 6d.; cloth boards, 3s. 6d.
7. **Rome, Outlines of the History of:** from the Earliest Period to the Christian Era and the Commencement of the Decline of the Empire. By EDWARD LEVIEN, of Balliol College, Oxford. Map, 2s. 6d.; cl. bds. 3s. 6d.
9. **Chronology of History, Art, Literature, and Progress,** from the Creation of the World to the Conclusion of the Franco-German War. The Continuation by W. D. HAMILTON, F.S.A. 3s.; cloth boards, 3s. 6d.
50. **Dates and Events in English History,** for the use of Candidates in Public and Private Examinations. By the Rev. E. RAND. 1s.

ENGLISH LANGUAGE AND MISCELLANEOUS.

11. **Grammar of the English Tongue,** Spoken and Written. With an Introduction to the Study of Comparative Philology. By HYDE CLARKE, D.C.L. Fourth Edition. 1s. 6d.
11*. **Philology:** Handbook of the Comparative Philology of English, Anglo-Saxon, Frisian, Flemish or Dutch, Low or Platt Dutch, High Dutch or German, Danish, Swedish, Icelandic, Latin, Italian, French, Spanish, and Portuguese Tongues. By HYDE CLARKE, D.C.L. 1s.
12. **Dictionary of the English Language,** as Spoken and Written. Containing above 100,000 Words. By HYDE CLARKE, D.C.L. 3s. 6d.; cloth boards, 4s. 6d.; complete with the GRAMMAR, cloth bds., 5s. 6d.
48. **Composition and Punctuation,** familiarly Explained for those who have neglected the Study of Grammar. By JUSTIN BRENAN. 17th Edition. 1s. 6d.
49. **Derivative Spelling-Book:** Giving the Origin of Every Word from the Greek, Latin, Saxon, German, Teutonic, Dutch, French, Spanish, and other Languages; with their present Acceptation and Pronunciation. By J. ROWBOTHAM, F.R.A.S. Improved Edition. 1s. 6d.
51. **The Art of Extempore Speaking:** Hints for the Pulpit, the Senate, and the Bar. By M. BAUTAIN, Vicar-General and Professor at the Sorbonne. Translated from the French. 7th Edition, carefully corrected. 2s. 6d.
52. **Mining and Quarrying,** with the Sciences connected therewith. First Book of, for Schools. By J. H. COLLINS, F.G.S., Lecturer to the Miners' Association of Cornwall and Devon. 1s.
53. **Places and Facts in Political and Physical Geography,** for Candidates in Examinations. By the Rev. EDGAR RAND, B.A. 1s.
54. **Analytical Chemistry,** Qualitative and Quantitative, a Course of. To which is prefixed, a Brief Treatise upon Modern Chemical Nomenclature and Notation. By WM. W. PINK and GEORGE E. WEBSTER. 2s.

THE SCHOOL MANAGERS' SERIES OF READING BOOKS,
Adapted to the Requirements of the New Code. Edited by the Rev. A. R. GRANT, Rector of Hitcham, and Honorary Canon of Ely; formerly H.M. Inspector of Schools.

INTRODUCTORY PRIMER, 3d.

	s. d.		s. d.
FIRST STANDARD	0 6	FOURTH STANDARD	1 2
SECOND ,,	0 10	FIFTH ,,	1 6
THIRD ,,	1 0	SIXTH ,,	1 6

LESSONS FROM THE BIBLE. Part I. Old Testament. 1s.
LESSONS FROM THE BIBLE. Part II. New Testament, to which is added THE GEOGRAPHY OF THE BIBLE, for very young Children. By Rev. C. THORNTON FORSTER. 1s. 2d. *₊* Or the Two Parts in One Volume. 2s.

14   WEALE'S EDUCATIONAL AND CLASSICAL SERIES.

FRENCH.

24. **French Grammar.** With Complete and Concise Rules on the Genders of French Nouns. By G. L. STRAUSS, Ph.D. 1s. 6d.
25. **French-English Dictionary.** Comprising a large number of New Terms used in Engineering, Mining, &c. By ALFRED ELWES. 1s. 6d.
26. **English-French Dictionary.** By ALFRED ELWES. 2s.
25,26. **French Dictionary** (as above). Complete, in One Vol., 3s.; cloth boards, 3s. 6d. *⁎* Or with the GRAMMAR, cloth boards, 4s. 6d.
47. **French and English Phrase Book:** containing Introductory Lessons, with Translations, several Vocabularies of Words, a Collection of suitable Phrases, and Easy Familiar Dialogues. 1s. 6d.

GERMAN.

39. **German Grammar.** Adapted for English Students, from Heyse's Theoretical and Practical Grammar, by Dr. G. L. STRAUSS. 1s.
40. **German Reader:** A Series of Extracts, carefully culled from the most approved Authors of Germany; with Notes, Philological and Explanatory. By G. L. STRAUSS, Ph.D. 1s.
41-43. **German Triglot Dictionary.** By NICHOLAS ESTERHAZY S. A. HAMILTON. In Three Parts. Part I. German-French-English. Part II. English-German-French. Part III. French-German-English. 3s., or cloth boards, 4s.
41-43 & 39. **German Triglot Dictionary** (as above), together with German Grammar (No. 39), in One Volume, cloth boards, 5s.

ITALIAN.

27. **Italian Grammar,** arranged in Twenty Lessons, with a Course of Exercises. By ALFRED ELWES. 1s. 6d.
28. **Italian Triglot Dictionary,** wherein the Genders of all the Italian and French Nouns are carefully noted down. By ALFRED ELWES. Vol. 1. Italian-English-French. 2s. 6d.
30. **Italian Triglot Dictionary.** By A. ELWES. Vol. 2. English-French-Italian. 2s. 6d.
32. **Italian Triglot Dictionary.** By ALFRED ELWES. Vol. 3. French-Italian-English. 2s. 6d.
28,30, **Italian Triglot Dictionary** (as above). In One Vol., 7s. 6d.
32.   Cloth boards.

SPANISH AND PORTUGUESE.

34. **Spanish Grammar,** in a Simple and Practical Form. With a Course of Exercises. By ALFRED ELWES. 1s. 6d.
35. **Spanish-English and English-Spanish Dictionary.** Including a large number of Technical Terms used in Mining, Engineering, &c., with the proper Accents and the Gender of every Noun. By ALFRED ELWES. 4s.; cloth boards, 5s. *⁎* Or with the GRAMMAR, cloth boards, 6s.
55. **Portuguese Grammar,** in a Simple and Practical Form. With a Course of Exercises. By ALFRED ELWES. 1s. 6d.
56. **Portuguese-English and English-Portuguese Dictionary,** with the Genders of each Noun. By ALFRED ELWES.
[*In preparation.*

HEBREW.

46*. **Hebrew Grammar.** By Dr. BRESSLAU. 1s. 6d.
44. **Hebrew and English Dictionary,** Biblical and Rabbinical; containing the Hebrew and Chaldee Roots of the Old Testament Post-Rabbinical Writings. By Dr. BRESSLAU. 6s.
46. **English and Hebrew Dictionary.** By Dr. BRESSLAU. 3s.
44,46. **Hebrew Dictionary** (as above), in Two Vols., complete, with 46*. the GRAMMAR, cloth boards, 12s.

LONDON: CROSBY LOCKWOOD AND CO.,

LATIN.

19. **Latin Grammar.** Containing the Inflections and Elementary Principles of Translation and Construction. By the Rev. THOMAS GOODWIN, M.A., Head Master of the Greenwich Proprietary School. 1s.
20. **Latin-English Dictionary.** By the Rev. THOMAS GOODWIN, M.A. 2s.
22. **English-Latin Dictionary;** together with an Appendix of French and Italian Words which have their origin from the Latin. By the Rev. THOMAS GOODWIN, M.A. 1s. 6d.
20,22. **Latin Dictionary** (as above). Complete in One Vol., 3s. 6d.; cloth boards, 4s. 6d. *⁎* Or with the GRAMMAR, cloth boards, 5s. 6d.

LATIN CLASSICS. With Explanatory Notes in English.
1. **Latin Delectus.** Containing Extracts from Classical Authors, with Genealogical Vocabularies and Explanatory Notes, by H. YOUNG. 1s. 6d.
2. **Cæsaris Commentarii** de Bello Gallico. Notes, and a Geographical Register for the Use of Schools, by H. YOUNG. 2s.
3. **Cornelius Nepos.** With Notes. By H. YOUNG. 1s.
4. **Virgilii Maronis Bucolica et Georgica.** With Notes on the Bucolics by W. RUSHTON, M.A., and on the Georgics by H. YOUNG. 1s. 6d.
5. **Virgilii Maronis Æneis.** With Notes, Critical and Explanatory, by H. YOUNG. New Edition, revised and improved. With copious Additional Notes by Rev. T. H. L. LEARY, D.C.L., formerly Scholar of Brasenose College, Oxford. 3s.
5⁕ ——— Part 1. Books i.—vi., 1s. 6d.
5⁕⁕ ——— Part 2. Books vii.—xii., 2s.
6. **Horace;** Odes, Epode, and Carmen Sæculare. Notes by H. YOUNG. 1s. 6d.
7. **Horace;** Satires, Epistles, and Ars Poetica. Notes by W. BROWNRIGG SMITH, M.A., F.R.G.S. 1s. 6d.
8. **Sallustii Crispi Catalina et Bellum Jugurthinum.** Notes, Critical and Explanatory, by W. M. DONNE, B.A., Trin. Coll., Cam. 1s. 6d.
9. **Terentii Andria et Heautontimorumenos.** With Notes, Critical and Explanatory, by the Rev. JAMES DAVIES, M.A. 1s. 6d.
10. **Terentii Adelphi, Hecyra, Phormio.** Edited, with Notes, Critical and Explanatory, by the Rev. JAMES DAVIES, M.A. 2s.
11. **Terentii Eunuchus, Comœdia.** Notes, by Rev. J. DAVIES, M.A. 1s. 6d.
12. **Ciceronis Oratio pro Sexto Roscio Amerino.** Edited, with an Introduction, Analysis, and Notes, Explanatory and Critical, by the Rev. JAMES DAVIES, M.A. 1s.
13. **Ciceronis Orationes in Catilinam, Verrem, et pro Archia.** With Introduction, Analysis, and Notes, Explanatory and Critical, by Rev. T. H. L. LEARY, D.C.L. formerly Scholar of Brasenose College, Oxford. 1s. 6d.
14. **Ciceronis Cato Major, Lælius, Brutus, sive de Senectute, de Amicitia, de Claris Oratoribus Dialogi.** With Notes by W. BROWNRIGG SMITH, M.A., F.R.G.S. 2s.
16. **Livy:** History of Rome. Notes by H. YOUNG and W. B. SMITH, M.A. Part 1. Books i., ii., 1s. 6d.
16⁕. ——— Part 2. Books iii., iv., v., 1s. 6d.
17. ——— Part 3. Books xxi., xxii., 1s. 6d.
19. **Latin Verse Selections,** from Catullus, Tibullus, Propertius, and Ovid. Notes by W. B. DONNE, M.A., Trinity College, Cambridge. 2s.
20. **Latin Prose Selections,** from Varro, Columella, Vitruvius, Seneca, Quintilian, Florus, Velleius Paterculus, Valerius Maximus Suetonius, Apuleius, &c. Notes by W. B. DONNE, M.A. 2s.
21. **Juvenalis Satiræ.** With Prolegomena and Notes by T. H. S. ESCOTT, B.A., Lecturer on Logic at King's College, London. 2s.

GREEK.

14. **Greek Grammar**, in accordance with the Principles and Philological Researches of the most eminent Scholars of our own day. By HANS CLAUDE HAMILTON. 1s. 6d.
15,17. **Greek Lexicon.** Containing all the Words in General Use, with their Significations, Inflections, and Doubtful Quantities. By HENRY R. HAMILTON. Vol. 1. Greek-English, 2s. 6d.; Vol. 2. English-Greek, 2s. Or the Two Vols. in One, 4s. 6d.; cloth boards, 5s.
14,15. **Greek Lexicon** (as above). Complete, with the GRAMMAR, in 17. One Vol., cloth boards, 6s.

GREEK CLASSICS. With Explanatory Notes in English.

1. **Greek Delectus.** Containing Extracts from Classical Authors, with Genealogical Vocabularies and Explanatory Notes, by H. YOUNG. New Edition, with an improved and enlarged Supplementary Vocabulary, by JOHN HUTCHISON, M.A., of the High School, Glasgow. 1s. 6d.
2, 3. **Xenophon's Anabasis;** or, The Retreat of the Ten Thousand. Notes and a Geographical Register, by H. YOUNG. Part 1. Books i. to iii., 1s. Part 2. Books iv. to vii., 1s.
4. **Lucian's Select Dialogues.** The Text carefully revised, with Grammatical and Explanatory Notes, by H. YOUNG. 1s. 6d.
5-12. **Homer, The Works of.** According to the Text of BAEUMLEIN. With Notes, Critical and Explanatory, drawn from the best and latest Authorities, with Preliminary Observations and Appendices, by T. H. L. LEARY, M.A., D.C.L.

THE ILIAD: Part 1. Books i. to vi., 1s. 6d. | Part 3. Books xiii. to xviii., 1s. 6d.
Part 2. Books vii. to xii., 1s. 6d. | Part 4. Books xix. to xxiv., 1s. 6d.
THE ODYSSEY: Part 1. Books i. to vi., 1s. 6d | Part 3. Books xiii. to xviii., 1s. 6d.
Part 2. Books vii. to xii., 1s. 6d. | Part 4. Books xix. to xxiv., and Hymns, 2s.

13. **Plato's Dialogues:** The Apology of Socrates, the Crito, and the Phædo. From the Text of C. F. HERMANN. Edited with Notes, Critical and Explanatory, by the Rev. JAMES DAVIES, M.A. 2s.
14-17. **Herodotus, The History of,** chiefly after the Text of GAISFORD. With Preliminary Observations and Appendices, and Notes, Critical and Explanatory, by T. H. L. LEARY, M.A., D.C.L.
Part 1. Books i., ii. (The Clio and Euterpe), 2s.
Part 2. Books iii., iv. (The Thalia and Melpomene), 2s.
Part 3. Books v.-vii. (The Terpsichore, Erato, and Polymnia), 2s.
Part 4. Books viii., ix. (The Urania and Calliope) and Index, 1s. 6d.
18. **Sophocles:** Œdipus Tyrannus. Notes by H. YOUNG. 1s.
20. **Sophocles:** Antigone. From the Text of DINDORF. Notes, Critical and Explanatory, by the Rev. JOHN MILNER, B.A. 2s.
23. **Euripides:** Hecuba and Medea. Chiefly from the Text of DINDORF. With Notes, Critical and Explanatory, by W. BROWNRIGG SMITH, M.A., F.R.G.S. 1s. 6d.
26. **Euripides:** Alcestis. Chiefly from the Text of DINDORF. With Notes, Critical and Explanatory, by JOHN MILNER, B.A. 1s. 6d.
30. **Æschylus:** Prometheus Vinctus: The Prometheus Bound. From the Text of DINDORF. Edited, with English Notes, Critical and Explanatory, by the Rev. JAMES DAVIES, M.A. 1s.
32. **Æschylus:** Septem Contra Thebes: The Seven against Thebes. From the Text of DINDORF. Edited, with English Notes, Critical and Explanatory, by the Rev. JAMES DAVIES, M.A. 1s.
40. **Aristophanes:** Acharnians. Chiefly from the Text of C. H. WEISE. With Notes, by C. S. T. TOWNSHEND, M.A. 1s. 6d.
41. **Thucydides:** History of the Peloponnesian War. Notes by H. YOUNG. Book 1. 1s.
42. **Xenophon's Panegyric on Agesilaus.** Notes and Introduction by LL. F. W. JEWITT. 1s. 6d.
43. **Demosthenes.** The Oration on the Crown and the Philippics. With English Notes. By Rev. T. H. L. LEARY, D.C.L., formerly Scholar of Brasenose College, Oxford. 1s. 6d.

LONDON, *October*, 1882.

A Catalogue of Books

INCLUDING MANY NEW AND STANDARD WORKS IN

ENGINEERING, ARCHITECTURE, AGRICULTURE, MATHEMATICS, MECHANICS, SCIENCE, ETC.

PUBLISHED BY

CROSBY LOCKWOOD & CO.,

7, STATIONERS'-HALL COURT, LUDGATE HILL, E.C.

ENGINEERING, SURVEYING, ETC.

Humber's New Work on Water-Supply.

A COMPREHENSIVE TREATISE on the WATER-SUPPLY of CITIES and TOWNS. By WILLIAM HUMBER, A.-M. Inst. C.E., and M. Inst. M.E. Illustrated with 50 Double Plates, 1 Single Plate, Coloured Frontispiece, and upwards of 250 Woodcuts, and containing 400 pages of Text. Imp. 4to, 6*l*. 6*s*. elegantly and substantially half-bound in morocco.

List of Contents:—

I. Historical Sketch of some of the means that have been adopted for the Supply of Water to Cities and Towns.—II. Water and the Foreign Matter usually associated with it.—III. Rainfall and Evaporation.—IV. Springs and the water-bearing formations of various districts.—V. Measurement and Estimation of the Flow of Water.—VI. On the Selection of the Source of Supply.—VII. Wells.—VIII. Reservoirs.—IX. The Purification of Water.—X. Pumps.—XI. Pumping Machinery.—XII. Conduits.—XIII. Distribution of Water.—XIV. Meters, Service Pipes, and House Fittings.—XV. The Law and Economy of Water Works.—XVI. Constant and Intermittent Supply.—XVII. Description of Plates.—Appendices, giving Tables of Rates of Supply, Velocities, &c. &c., together with Specifications of several Works illustrated, among which will be found:—Aberdeen, Bideford, Canterbury, Dundee, Halifax, Lambeth, Rotherham, Dublin, and others.

"The most systematic and valuable work upon water supply hitherto produced in English, or in any other language . . . . Mr. Humber's work is characterised almost throughout by an exhaustiveness much more distinctive of French and German than of English technical treatises."—*Engineer.*

Humber's Great Work on Bridge Construction.

A COMPLETE and PRACTICAL TREATISE on CAST and WROUGHT-IRON BRIDGE CONSTRUCTION, including Iron Foundations. In Three Parts—Theoretical, Practical, and Descriptive. By WILLIAM HUMBER, A.-M. Inst. C.E., and M. Inst. M.E. Third Edition, with 115 Double Plates. In 2 vols. imp. 4to, 6*l*. 16*s*. 6*d*. half-bound in morocco.

"A book—and particularly a large and costly treatise like Mr. Humber's—which has reached its third edition may certainly be said to have established its own reputation."—*Engineering.*

WORKS IN ENGINEERING, SURVEYING, ETC.,

Humber's Modern Engineering.
A RECORD of the PROGRESS of MODERN ENGINEERING. First Series. Comprising Civil, Mechanical, Marine, Hydraulic, Railway, Bridge, and other Engineering Works, &c. By WILLIAM HUMBER, A.-M. Inst. C. E., &c. Imp. 4to, with 36 Double Plates, drawn to a large scale, and Portrait of John Hawkshaw, C.E., F.R.S., &c., and descriptive Letter-press, Specifications, &c. 3*l.* 3*s.* half morocco.
List of the Plates and Diagrams.
Victoria Station and Roof, L. B. & S. C. R. (8 plates); Southport Pier (2 plates); Victoria Station and Roof, L. C. & D. and G. W. R. (6 plates); Roof of Cremorne Music Hall; Bridge over G. N. Railway; Roof of Station, Dutch Rhenish Rail (2 plates); Bridge over the Thames, West London Extension Railway (5 plates); Armour Plates; Suspension Bridge, Thames (4 plates); The Allen Engine; Suspension Bridge, Avon (3 plates); Underground Railway (3 plates).
"Handsomely lithographed and printed. It will find favour with many who desire to preserve in a permanent form copies of the plans and specifications prepared for the guidance of the contractors for many important engineering works."—*Engineer.*

HUMBER'S RECORD OF MODERN ENGINEERING. Second Series. Imp. 4to, with 36 Double Plates, Portrait of Robert Stephenson, C.E., &c., and descriptive Letterpress, Specifications, &c. 3*l.* 3*s.* half morocco.
List of the Plates and Diagrams.
Birkenhead Docks, Low Water Basin (15 plates); Charing Cross Station Roof, C. C. Railway (3 plates); Digswell Viaduct, G. N. Railway; Robbery Wood Viaduct, G. N. Railway; Iron Permanent Way; Clydach Viaduct, Merthyr, Tredegar, and Abergavenny Railway; Ebbw Viaduct, Merthyr, Tredegar, and Abergavenny Railway; College Wood Viaduct, Cornwall Railway; Dublin Winter Palace Roof (3 plates); Bridge over the Thames, L. C. and D. Railway (6 plates); Albert Harbour, Greenock (4 plates).

HUMBER'S RECORD OF MODERN ENGINEERING. Third Series. Imp. 4to, with 40 Double Plates, Portrait of J. R. M'Clean, Esq., late Pres. Inst. C. E., and descriptive Letterpress, Specifications, &c. 3*l.* 3*s.* half morocco.
List of the Plates and Diagrams.
MAIN DRAINAGE, METROPOLIS.—*North Side.*—Map showing Interception of Sewers; Middle Level Sewer (2 plates); Outfall Sewer, Bridge over River Lea (3 plates); Outfall Sewer, Bridge over Marsh Lane, North Woolwich Railway, and Bow and Barking Railway Junction; Outfall Sewer, Bridge over Bow and Barking Railway (3 plates); Outfall Sewer, Bridge over East London Waterworks' Feeder (2 plates); Outfall Sewer, Reservoir (2 plates); Outfall Sewer, Tumbling Bay and Outlet; Outfall Sewer, Penstocks. *South Side.*—Outfall Sewer, Bermondsey Branch (2 plates); Outfall Sewer, Reservoir and Outlet (4 plates); Outfall Sewer, Filth Hoist; Sections of Sewers (North and South Sides).
THAMES EMBANKMENT.—Section of River Wall; Steamboat Pier, Westminster (2 plates); Landing Stairs between Charing Cross and Waterloo Bridges; York Gate (2 plates); Overflow and Outlet at Savoy Street Sewer (3 plates); Steamboat Pier, Waterloo Bridge (3 plates); Junction of Sewers, Plans and Sections; Gullies, Plans and Sections; Rolling Stock; Granite and Iron Forts.

HUMBER'S RECORD OF MODERN ENGINEERING. Fourth Series. Imp. 4to, with 36 Double Plates, Portrait of John Fowler, Esq., late Pres. Inst. C.E., and descriptive Letterpress, Specifications, &c. 3*l.* 3*s.* half morocco.
List of the Plates and Diagrams.
Abbey Mills Pumping Station, Main Drainage, Metropolis (4 plates); Barrow Docks (5 plates); Manquis Viaduct, Santiago and Valparaiso Railway (2 plates); Adam's Locomotive, St. Helen's Canal Railway (2 plates); Cannon Street Station Roof, Charing Cross Railway (3 plates); Road Bridge over the River Moka (2 plates). Telegraphic Apparatus for Mesopotamia; Viaduct over the River Wye, Midland Railway (3 plates); St. German's Viaduct, Cornwall Railway (2 plates); Wrought-Iron Cylinder for Diving Bell; Millwall Docks (6 plates); Milroy's Patent Excavator, Metropolitan District Railway (6 plates); Harbours, Ports, and Breakwaters (3 plates).

Strains, Formulæ & Diagrams for Calculation of.

A HANDY BOOK for the CALCULATION of STRAINS in GIRDERS and SIMILAR STRUCTURES, and their STRENGTH; consisting of Formulæ and Corresponding Diagrams, with numerous Details for Practical Application, &c. By WILLIAM HUMBER, A.-M. Inst. C.E., &c. Third Edition. With nearly 100 Woodcuts and 3 Plates, Crown 8vo, 7s. 6d. cloth.

"The system of employing diagrams as a substitute for complex computations is one justly coming into great favour, and in that respect Mr. Humber's volume is fully up to the times."—*Engineering.*

Strains.

THE STRAINS ON STRUCTURES OF IRONWORK; with Practical Remarks on Iron Construction. By F. W. SHEILDS, M. Inst. C.E. Second Edition, with 5 Plates. Royal 8vo, 5s. cloth.

"The student cannot find a better book on this subject than Mr. Sheilds'."—*Engineer.*

Barlow on the Strength of Materials, enlarged.

A TREATISE ON THE STRENGTH OF MATERIALS, with Rules for application in Architecture, the Construction of Suspension Bridges, Railways, &c.; and an Appendix on the Power of Locomotive Engines, and the effect of Inclined Planes and Gradients. By PETER BARLOW, F.R.S. A New Edition, revised by his Sons, P. W. BARLOW, F.R.S., and W. H. BARLOW, F.R.S. The whole arranged and edited by W. HUMBER, A-M. Inst. C.E. 8vo, 400 pp., with 19 large Plates, 18s. cloth.

"The standard treatise upon this particular subject."—*Engineer.*

Strength of Cast Iron, &c.

A PRACTICAL ESSAY on the STRENGTH of CAST IRON and OTHER METALS. By T. TREDGOLD, C.E. 5th Edition. To which are added, Experimental Researches on the Strength, &c., of Cast Iron, by E. HODGKINSON, F.R.S. 8vo, 12s. cloth.

_{*} HODGKINSON'S RESEARCHES, separate, price 6s.

Hydraulics.

HYDRAULIC TABLES, CO-EFFICIENTS, and FORMULÆ for finding the Discharge of Water from Orifices, Notches, Weirs, Pipes, and Rivers. With New Formulæ, Tables, and General Information on Rain-fall, Catchment-Basins, Drainage, Sewerage, Water Supply for Towns and Mill Power. By JOHN NEVILLE, Civil Engineer, M.R.I.A. Third Edition, carefully revised, with considerable Additions. Numerous Illustrations. Cr. 8vo, 14s. cloth.

"Undoubtedly an exceedingly useful and elaborate compilation."—*Iron.*
"Alike valuable to students and engineers in practice."—*Mining Journal.*

River Engineering.

RIVER BARS: Notes on the Causes of their Formation, and on their Treatment by 'Induced Tidal Scour,' with a Description of the Successful Reduction by this Method of the Bar at Dublin. By I. J. MANN, Assistant Engineer to the Dublin Port and Docks Board. With Illustrations. Royal 8vo. 7s. 6d. cloth.

Hydraulics.

HYDRAULIC MANUAL. Consisting of Working Tables and Explanatory Text. Intended as a Guide in Hydraulic Calculations and Field Operations. By LOWIS D'A. JACKSON. Fourth Edition. Rewritten and Enlarged. Large Crown 8vo. [*In the press.*

4 WORKS IN ENGINEERING, SURVEYING, ETC.,

Levelling.

A TREATISE on the PRINCIPLES and PRACTICE of LEVELLING; showing its Application to Purposes of Railway and Civil Engineering, in the Construction of Roads; with Mr. TELFORD'S Rules for the same. By FREDERICK W. SIMMS, F.G.S., M. Inst. C.E. Sixth Edition, very carefully revised, with the addition of Mr. LAW'S Practical Examples for Setting out Railway Curves, and Mr. TRAUTWINE'S Field Practice of Laying out Circular Curves. With 7 Plates and numerous Woodcuts. 8vo, 8s. 6d. cloth. *** TRAUTWINE on Curves, separate, 5s.
"The text-book on levelling in most of our engineering schools and colleges."—*Engineer.*

Practical Tunnelling.

PRACTICAL TUNNELLING: Explaining in detail the Setting out of the Works, Shaft-sinking and Heading-Driving, Ranging the Lines and Levelling under Ground, Sub-Excavating, Timbering, and the Construction of the Brickwork of Tunnels with the amount of labour required for, and the Cost of, the various portions of the work. By F. W. SIMMS, M. Inst. C.E. Third Edition, Revised and Extended. By D. KINNEAR CLARK, M.I.C.E. Imp. 8vo, with 21 Folding Plates and numerous Wood Engravings, 30s. cloth.
"It has been regarded from the first as a text-book of the subject. . . . Mr. Clark has added immensely to the value of the book."—*Engineer.*

Steam.

STEAM AND THE STEAM ENGINE, Stationary and Portable. Being an Extension of Sewell's Treatise on Steam. By D. KINNEAR CLARK, M.I.C.E. Second Edition. 12mo, 4s. cloth.

Civil and Hydraulic Engineering.

CIVIL ENGINEERING. By HENRY LAW, M. Inst. C.E. Including a Treatise on Hydraulic Engineering, by GEORGE R. BURNELL, M.I.C.E. Sixth Edition, Revised, with large additions on Recent Practice in Civil Engineering, by D. KINNEAR CLARK, M. Inst. C.E. 12mo, 7s. 6d., cloth boards.

Gas-Lighting.

COMMON SENSE FOR GAS-USERS: a Catechism of Gas-Lighting for Householders, Gasfitters, Millowners, Architects, Engineers, &c. By R. WILSON, C.E. 2nd Edition. Cr. 8vo, 2s. 6d.

Bridge Construction in Masonry, Timber, & Iron.

EXAMPLES OF BRIDGE AND VIADUCT CONSTRUCTION IN MASONRY, TIMBER, AND IRON; consisting of 46 Plates from the Contract Drawings or Admeasurement of select Works. By W. DAVIS HASKOLL, C.E. Second Edition, with the addition of 554 Estimates, and the Practice of Setting out Works, with 6 pages of Diagrams. Imp. 4to, 2l. 12s. 6d. half-morocco.
"A work of the present nature by a man of Mr. Haskoll's experience, must prove invaluable. The tables of estimates considerably enhance its value."—*Engineering.*

Earthwork.

EARTHWORK TABLES, showing the Contents in Cubic Yards of Embankments, Cuttings, &c., of Heights or Depths up to an average of 80 feet. By JOSEPH BROADBENT, C.E., and FRANCIS CAMPIN, C.E. Cr. 8vo, oblong, 5s. cloth.

Tramways and their Working.

TRAMWAYS: their CONSTRUCTION and WORKING. With Special Reference to the Tramways of the United Kingdom. By D. KINNEAR CLARK, M.I.C.E. SUPPLEMENTARY VOLUME; recording the Progress recently made in the Design and Construction of Tramways, and in the Means of Locomotion by Mechanical Power. With Wood Engravings. 8vo, 12s. cloth.

Tramways and their Working.

TRAMWAYS: their CONSTRUCTION and WORKING. By D. KINNEAR CLARK, M. Inst. C. E. With Wood Engravings, and thirteen folding Plates. THE COMPLETE WORK, in 2 vols., Large Crown 8vo, 30s. cloth.

"All interested in tramways must refer to it, as all railway engineers have turned to the author's work 'Railway Machinery.'"—*The Engineer.*

Pioneer Engineering.

PIONEER ENGINEERING. A Treatise on the Engineering Operations connected with the Settlement of Waste Lands in New Countries. By EDWARD DOBSON, A.I.C.E. With Plates and Wood Engravings. Revised Edition. 12mo, 5s. cloth.

"A workmanlike production, and one without possession of which no man should start to encounter the duties of a pioneer engineer."—*Athenæum.*

Steam Engine.

TEXT-BOOK ON THE STEAM ENGINE. By T. M. GOODEVE, M.A., Barrister-at-Law, Author of "The Principles of Mechanics," "The Elements of Mechanism," &c. Fourth Edition. With numerous Illustrations. Crown 8vo, 6s. cloth.

"Mr. Goodeve's text-book is a work of which every young engineer should possess himself."—*Mining Journal.*

Steam.

THE SAFE USE OF STEAM: containing Rules for Unprofessional Steam Users. By an ENGINEER. 4th Edition. Sewed, 6d.

"If steam-users would but learn this little book by heart, boiler explosions would become sensations by their rarity."—*English Mechanic.*

Mechanical Engineering.

DETAILS OF MACHINERY: Comprising Instructions for the Execution of various Works in Iron, in the Fitting-Shop, Foundry, and Boiler-Yard. Arranged expressly for the use of Draughtsmen, Students, and Foremen Engineers. By FRANCIS CAMPIN, C.E. 12mo. 3s. 6d. cloth. [*Just published.*

Mechanical Engineering.

MECHANICAL ENGINEERING: Comprising Metallurgy, Moulding, Casting, Forging, Tools, Workshop Machinery, Manufacture of the Steam Engine, &c. By F. CAMPIN, C.E. 3s. cloth.

Works of Construction.

MATERIALS AND CONSTRUCTION: a Theoretical and Practical Treatise on the Strains, Designing, and Erection of Works of Construction. By F. CAMPIN, C.E. 12mo. 3s. 6d. cl. brds.

Iron Bridges, Girders, Roofs, &c.

A TREATISE ON THE APPLICATION OF IRON TO THE CONSTRUCTION OF BRIDGES, GIRDERS, ROOFS, AND OTHER WORKS. By F. CAMPIN, C.E. 12mo, 3s.

Boiler Construction.

THE MECHANICAL ENGINEER'S OFFICE BOOK: Boiler Construction. By NELSON FOLEY, Cardiff, late Assistant Manager Palmer's Engine Works, Jarrow. With 29 full-page Lithographic Diagrams. Folio 21s. half-bound.

Oblique Arches.

A PRACTICAL TREATISE ON THE CONSTRUCTION of OBLIQUE ARCHES. By JOHN HART. 3rd Ed. Imp. 8vo, 8s. cloth.

Oblique Bridges.

A PRACTICAL and THEORETICAL ESSAY on OBLIQUE BRIDGES, with 13 large Plates. By the late GEO. WATSON BUCK, M.I.C.E. Third Edition, revised by his Son, J. H. WATSON BUCK, M.I.C.E.; and with the addition of Description to Diagrams for Facilitating the Construction of Oblique Bridges, by W. H. BARLOW, M.I.C.E. Royal 8vo, 12s. cloth.

"The standard text book for all engineers regarding skew arches."—*Engineer*.

Gas and Gasworks.

THE CONSTRUCTION OF GASWORKS AND THE MANUFACTURE AND DISTRIBUTION OF COAL-GAS. Originally written by S. HUGHES, C.E. Sixth Edition. Re-written and enlarged, by W. RICHARDS, C.E. 12mo, 5s. cloth.

Waterworks for Cities and Towns.

WATERWORKS for the SUPPLY of CITIES and TOWNS, with a Description of the Principal Geological Formations of England as influencing Supplies of Water. By S. HUGHES. 4s. 6d. cloth.

Locomotive-Engine Driving.

LOCOMOTIVE-ENGINE DRIVING; a Practical Manual for Engineers in charge of Locomotive Engines. By MICHAEL REYNOLDS, M.S.E. Fifth Edition. Comprising A KEY TO THE LOCOMOTIVE ENGINE. With Illustrations. Cr. 8vo, 4s. 6d. cl.

"Mr. Reynolds has supplied a want, and has supplied it well."—*Engineer*.

The Engineer, Fireman, and Engine-Boy.

THE MODEL LOCOMOTIVE ENGINEER, FIREMAN, AND ENGINE-BOY. By M. REYNOLDS. Crown 8vo, 4s. 6d.

Stationary Engine Driving.

STATIONARY ENGINE DRIVING. A Practical Manual for Engineers in Charge of Stationary Engines. By MICHAEL REYNOLDS. Second Edition, Revised and Enlarged. With Plates and Woodcuts. Crown 8vo, 4s. 6d. cloth. [*Just published*.

Engine-Driving Life.

ENGINE-DRIVING LIFE; or Stirring Adventures and Incidents in the Lives of Locomotive Engine-Drivers. By MICHAEL REYNOLDS. Crown 8vo, 2s. cloth.

Continuous Railway Brakes.

CONTINUOUS RAILWAY BRAKES. A Practical Treatise on the several Systems in Use in the United Kingdom; their Construction and Performance. With copious Illustrations and numerous Tables. By MICHAEL REYNOLDS. Large Crown 8vo, 9s. cloth.

[*Just published*.

Construction of Iron Beams, Pillars, &c.
IRON AND HEAT; exhibiting the Principles concerned in the construction of Iron Beams, Pillars, and Bridge Girders, and the Action of Heat in the Smelting Furnace. By J. ARMOUR, C.E. 3s.

Fire Engineering.
FIRES, FIRE-ENGINES, AND FIRE BRIGADES. With a History of Fire-Engines, their Construction, Use, and Management; Remarks on Fire-Proof Buildings, and the Preservation of Life from Fire; Statistics of the Fire Appliances in English Towns; Foreign Fire Systems; Hints on Fire Brigades, &c., &c. By CHARLES F. T. YOUNG, C.E. With numerous Illustrations, handsomely printed, 544 pp., demy 8vo, 1l. 4s. cloth.
"We can most heartily commend this book."—*Engineering.*
"Mr. Young's book on 'Fire Engines and Fire Brigades' contains a mass of information, which has been collected from a variety of sources. The subject is so intensely interesting and useful that it demands consideration."—*Building News.*

Trigonometrical Surveying.
AN OUTLINE OF THE METHOD OF CONDUCTING A TRIGONOMETRICAL SURVEY, for the Formation of Geographical and Topographical Maps and Plans, Military Reconnaissance, Levelling, &c., with the most useful Problems in Geodesy and Practical Astronomy. By LIEUT.-GEN. FROME, R.E., late Inspector-General of Fortifications. Fourth Edition, Enlarged, and partly Re-written. By CAPTAIN CHARLES WARREN, R.E. With 19 Plates and 115 Woodcuts, royal 8vo, 16s. cloth.

Tables of Curves.
TABLES OF TANGENTIAL ANGLES and MULTIPLES for setting out Curves from 5 to 200 Radius. By ALEXANDER BEAZELEY, M. Inst. C.E. Second Edition. Printed on 48 Cards, and sold in a cloth box, waistcoat-pocket size, 3s. 6d.
"Each table is printed on a small card, which, being placed on the theodolite, leaves the hands free to manipulate the instrument."—*Engineer.*
"Very handy; a man may know that all his day's work must fall on two of these cards, which he puts into his own card-case, and leaves the rest behind."—[*Athenæum.*

Engineering Fieldwork.
THE PRACTICE OF ENGINEERING FIELDWORK, applied to Land and Hydraulic, Hydrographic, and Submarine Surveying and Levelling. Second Edition, revised, with considerable additions, and a Supplement on WATERWORKS, SEWERS, SEWAGE, and IRRIGATION. By W. DAVIS HASKOLL, C.E. Numerous folding Plates. In 1 Vol., demy 8vo, 1l. 5s., cl. boards.

Large Tunnel Shafts.
THE CONSTRUCTION OF LARGE TUNNEL SHAFTS. A Practical and Theoretical Essay. By J. H. WATSON BUCK, M. Inst. C.E., Resident Engineer, London and North-Western Railway. Illustrated with Folding Plates. Royal 8vo, 12s. cloth.
"Many of the methods given are of extreme practical value to the mason, and the observations on the form of arch, the rules for ordering the stone, and the construction of the templates, will be found of considerable use. We commend the book to the engineering profession, and to all who have to build similar shafts."—*Building News.*
"Will be regarded by civil engineers as of the utmost value, and calculated to save much time and obviate many mistakes."—*Colliery Guardian.*

8   WORKS IN ENGINEERING, SURVEYING, ETC.,

Survey Practice.
AID TO SURVEY PRACTICE: for Reference in Surveying, Levelling, Setting-out and in Route Surveys of Travellers by Land and Sea. With Tables, Illustrations, and Records. By LOWIS D'A. JACKSON, A.-M.I.C.E. Author of "Hydraulic Manual and Statistics," &c. Large crown 8vo, 12s. 6d., cloth.
"Mr. Jackson has produced a valuable *vade-mecum* for the surveyor. We can recommend this book as containing an admirable supplement to the teaching of the accomplished surveyor."—*Athenæum*.
"A general text book was wanted, and we are able to speak with confidence of Mr. Jackson's treatise. . . . We cannot recommend to the student who knows something of the mathematical principles of the subject a better course than to fortify his practice in the field under a competent surveyor with a study of Mr. Jackson's useful manual. The field records illustrate every kind of survey, and will be found an essential aid to the student."—*Building News*.
"The author brings to his work a fortunate union of theory and practical experience which, aided by a clear and lucid style of writing, renders the book both a very useful one and very agreeable to read."—*Builder*.

Sanitary Work.
SANITARY WORK IN THE SMALLER TOWNS AND IN VILLAGES. Comprising:—1. Some of the more Common Forms of Nuisance and their Remedies; 2. Drainage; 3. Water Supply. By CHAS. SLAGG, Assoc. Inst. C.E. Crown 8vo, 3s. cloth.
"A very useful book, and may be safely recommended. The author has had practical experience in the works of which he treats."—*Builder*.

Locomotives.
LOCOMOTIVE ENGINES, A Rudimentary Treatise on. Comprising an Historical Sketch and Description of the Locomotive Engine. By G. D. DEMPSEY, C.E. With large additions treating of the MODERN LOCOMOTIVE, by D. KINNEAR CLARK, C.E., M.I.C.E., Author of "Tramways, their Construction and Working," &c., &c. With numerous Illustrations. 12mo. 3s. 6d. cloth boards.
"The student cannot fail to profit largely by adopting this as his preliminary textbook."—*Iron and Coal Trades Review*.
"Seems a model of what an elementary technical book should be."—*Academy*.

Fuels and their Economy.
FUEL, its Combustion and Economy; consisting of an Abridgment of "A Treatise on the Combustion of Coal and the Prevention of Smoke." By C. W. WILLIAMS, A.I.C.E. With extensive additions on Recent Practice in the Combustion and Economy of Fuel—Coal, Coke, Wood, Peat, Petroleum, &c.; by D. KINNEAR CLARK, C.E., M.I.C.E. Second Edition, revised. With numerous Illustrations. 12mo. 4s. cloth boards.
"Students should buy the book and read it, as one of the most complete and satisfactory treatises on the combustion and economy of fuel to be had."—*Engineer*.

Roads and Streets.
THE CONSTRUCTION OF ROADS AND STREETS. In Two Parts. I. The Art of Constructing Common Roads. By HENRY LAW, C.E. Revised and Condensed. II. Recent Practice in the Construction of Roads and Streets: including Pavements of Stone, Wood, and Asphalte. By D. KINNEAR CLARK, C.E., M.I.C.E. Second Edit., revised. 12mo, 5s. cloth.
"A book which every borough surveyor and engineer must possess, and which will be of considerable service to architects, builders, and property owners generally."—*Building News*.

Sewing Machine (The).

SEWING MACHINERY; being a Practical Manual of the Sewing Machine, comprising its History and Details of its Construction, with full Technical Directions for the Adjusting of Sewing Machines. By J. W. URQUHART, Author of "Electro Plating: a Practical Manual;" "Electric Light: its Production and Use." With Numerous Illustrations. 12mo, 2s. 6d. cloth.

Field-Book for Engineers.

THE ENGINEER'S, MINING SURVEYOR'S, and CONTRACTOR'S FIELD-BOOK. By W. DAVIS HASKOLL, C.E. Consisting of a Series of Tables, with Rules, Explanations of Systems, and Use of Theodolite for Traverse Surveying and Plotting the Work with minute accuracy by means of Straight Edge and Set Square only; Levelling with the Theodolite, Casting out and Reducing Levels to Datum, and Plotting Sections in the ordinary manner; Setting out Curves with the Theodolite by Tangential Angles and Multiples with Right and Left-hand Readings of the Instrument; Setting out Curves without Theodolite on the System of Tangential Angles by Sets of Tangents and Offsets; and Earthwork Tables to 80 feet deep, calculated for every 6 inches in depth. With numerous Woodcuts. 4th Edition, enlarged. Cr. 8vo. 12s. cloth.

"The book is very handy, and the author might have added that the separate tables of sines and tangents to every minute will make it useful for many other purposes, the genuine traverse tables existing all the same."—*Athenæum.*
"Cannot fail, from its portability and utility, to be extensively patronised by the engineering profession."—*Mining Journal.*

Earthwork, Measurement and Calculation of.

A MANUAL on EARTHWORK. By ALEX. J. S. GRAHAM, C.E., Resident Engineer, Forest of Dean Central Railway. With numerous Diagrams. 18mo, 2s. 6d. cloth.

"As a really handy book for reference, we know of no work equal to it; and the railway engineers and others employed in the measurement and calculation of earthwork will find a great amount of practical information very admirably arranged, and available for general or rough estimates, as well as for the more exact calculations required in the engineers' contractor's offices."—*Artizan.*

Drawing for Engineers, &c.

THE WORKMAN'S MANUAL OF ENGINEERING DRAWING. By JOHN MAXTON, Instructor in Engineering Drawing, Royal Naval College, Greenwich, formerly of R. S. N. A., South Kensington. Fourth Edition, carefully revised. With upwards of 300 Plates and Diagrams. 12mo, cloth, strongly bound, 4s.

"A copy of it should be kept for reference in every drawing office."—*Engineering.*
"Indispensable for teachers of engineering drawing."—*Mechanics' Magazine.*

Weale's Dictionary of Terms.

A DICTIONARY of TERMS used in ARCHITECTURE, BUILDING, ENGINEERING, MINING, METALLURGY, ARCHÆOLOGY, the FINE ARTS, &c. By JOHN WEALE. Fifth Edition, revised by ROBERT HUNT, F.R.S., Keeper of Mining Records, Editor of "Ure's Dictionary of Arts." 12mo, 6s. cl. bds.

"The best small technological dictionary in the language."—*Architect.*
"The absolute accuracy of a work of this character can only be judged of after extensive consultation, and from our examination it appears very correct and very complete."—*Mining Journal.*

WORKS IN MINING, METALLURGY, ETC.,

MINING, METALLURGY, ETC.

Coal and Iron.
THE COAL AND IRON INDUSTRIES OF THE UNITED KINGDOM : comprising a Description of the Coal Fields, and of the Principal Seams of Coal, with returns of their Produce and its Distribution, and Analyses of Special Varieties. Also, an Account of the occurrence of Iron Ores in Veins or Seams ; Analyses of each Variety ; and a History of the Rise and Progress of Pig Iron Manufacture since the year 1740, exhibiting the economies introduced in the Blast Furnaces for its Production and Improvement. By RICHARD MEADE, Assistant Keeper of Mining Records. With Maps of the Coal Fields and Ironstone Deposits of the United Kingdom. 8vo., £1 8s. cloth. [*Just published.*

Metalliferous Minerals and Mining.
A TREATISE ON METALLIFEROUS MINERALS AND MINING. By D.C. DAVIES, F.G.S., author of "A Treatise on Slate and Slate Quarrying." With numerous wood engravings. Second Edition, revised. Cr. 8vo. 12s. 6d. cloth.
" Without question, the most exhaustive and the most practically useful work we have seen ; the amount of information given is enormous, and it is given concisely and intelligibly."—*Mining Journal.*

Slate and Slate Quarrying.
A TREATISE ON SLATE AND SLATE QUARRYING, Scientific, Practical, and Commercial. By D. C. DAVIES, F.G.S., Mining Engineer, &c. With numerous Illustrations and Folding Plates. Second Edition, carefully revised. 12mo, 3s. 6d. cloth boards.
"Mr. Davies has written a useful and practical hand-book on an important industry, with all the conditions and details of which he appears familiar."—*Engineering.*

Metallurgy of Iron.
A TREATISE ON THE METALLURGY OF IRON : containing Outlines of the History of Iron Manufacture, Methods of Assay, and Analyses of Iron Ores, Processes of Manufacture of Iron and Steel, &c. By H. BAUERMAN, F.G.S. Fifth Edition, Revised and greatly Enlarged. With Numerous Illustrations, 12mo. 5s. 6d., cloth boards. [*Just published.*

Manual of Mining Tools.
MINING TOOLS. For the use of Mine Managers, Agents, Mining Students, &c. By WILLIAM MORGANS. Volume of Text. 12mo, 3s. With an Atlas of Plates, containing 235 Illustrations. 4to, 6s. Together, 9s. cloth boards.

Mining, Surveying and Valuing.
THE MINERAL SURVEYOR AND VALUER'S COMPLETE GUIDE, comprising a Treatise on Improved Mining Surveying, with new Traverse Tables ; and Descriptions of Improved Instruments ; also an Exposition of the Correct Principles of Laying out and Valuing Home and Foreign Iron and Coal Mineral Properties. By WILLIAM LINTERN, Mining and Civil Engineer. With four Plates of Diagrams, Plans, &c., 12mo, 4s. cloth.
*** The above, bound with THOMAN'S TABLES. (See page 20.) Price 7s. 6d. cloth.

Coal and Coal Mining.

COAL AND COAL MINING: a Rudimentary Treatise on. By WARINGTON W. SMYTH, M.A., F.R.S., &c., Chief Inspector of the Mines of the Crown. Fifth edition, revised and corrected. 12mo, with numerous Illustrations, 4s. cloth boards.

"Every portion of the volume appears to have been prepared with much care, and as an outline is given of every known coal-field in this and other countries, as well as of the two principal methods of working, the book will doubtless interest a very large number of readers."—*Mining Journal.*

Underground Pumping Machinery.

MINE DRAINAGE; being a Complete and Practical Treatise on Direct-Acting Underground Steam Pumping Machinery, with a Description of a large number of the best known Engines, their General Utility and the Special Sphere of their Action, the Mode of their Application, and their merits compared with other forms of Pumping Machinery. By STEPHEN MICHELL, Joint-Author of "The Cornish System of Mine Drainage." 8vo, 15s. cloth.

NAVAL ARCHITECTURE, NAVIGATION, ETC.

Pocket Book for Naval Architects & Shipbuilders.

THE NAVAL ARCHITECT'S AND SHIPBUILDER'S POCKET BOOK OF FORMULÆ, RULES, AND TABLES AND MARINE ENGINEER'S AND SURVEYOR'S HANDY BOOK OF REFERENCE. By CLEMENT MACKROW, M. Inst. N. A., Naval Draughtsman. Second Edition, revised. With numerous Diagrams. Fcap., 12s. 6d., strongly bound in leather.

"Should be used by all who are engaged in the construction or design of vessels."—*Engineer.*

"There is scarcely a subject on which a naval architect or shipbuilder can require to refresh his memory which will not be found within the covers of Mr. Mackrow's book."—*English Mechanic.*

"Mr. Mackrow has compressed an extraordinary amount of information into this useful volume."—*Athenæum.*

Grantham's Iron Ship-Building.

ON IRON SHIP-BUILDING; with Practical Examples and Details. Fifth Edition. Imp. 4to, boards, enlarged from 24 to 40 Plates (21 quite new), including the latest Examples. Together with separate Text, also considerably enlarged, 12mo, cloth limp. By JOHN GRANTHAM, M. Inst. C.E., &c. 2l. 2s. complete.

"Mr. Grantham's work is of great interest. It will, we are confident, command an extensive circulation among shipbuilders in general. By order of the Board of Admiralty, the work will form the text-book on which the examination in iron ship-building of candidates for promotion in the dockyards will be mainly based."—*Engineering.*

Pocket-Book for Marine Engineers.

A POCKET-BOOK OF USEFUL TABLES AND FORMULÆ FOR MARINE ENGINEERS. By FRANK PROCTOR, A.I.N.A. Third Edition. Royal 32mo, leather, gilt edges, 4s.

"A most useful companion to all marine engineers."—*United Service Gazette.*

"Scarcely anything required by a naval engineer appears to have been forgotten."—*Iron.*

Light-Houses.

EUROPEAN LIGHT-HOUSE SYSTEMS; being a Report of a Tour of Inspection made in 1873. By Major GEORGE H. ELLIOT, Corps of Engineers, U.S.A. Illustrated by 51 Engravings and 31 Woodcuts in the Text. 8vo, 21s. cloth.

Surveying (Land and Marine).

LAND AND MARINE SURVEYING, In Reference to the Preparation of Plans for Roads and Railways, Canals, Rivers, Towns' Water Supplies, Docks and Harbours; with Description and Use of Surveying Instruments. By W. DAVIS HASKOLL, C.E. With 14 folding Plates, and numerous Woodcuts. 8vo, 12s. 6d. cloth.

"A most useful and well arranged book for the aid of a student."—*Builder*.

"Of the utmost practical utility, and may be safely recommended to all students who aspire to become clean and expert surveyors."—*Mining Journal*.

Storms.

STORMS: their Nature, Classification, and Laws, with the Means of Predicting them by their Embodiments, the Clouds. By WILLIAM BLASIUS. Crown 8vo, 10s. 6d. cloth boards.

Rudimentary Navigation.

THE SAILOR'S SEA-BOOK: a Rudimentary Treatise on Navigation. By JAMES GREENWOOD, B.A. New and enlarged edition. By W. H. ROSSER. 12mo, 3s. cloth boards.

Mathematical and Nautical Tables.

MATHEMATICAL TABLES, for Trigonometrical, Astronomical and Nautical Calculations; to which is prefixed a Treatise on Logarithms. By HENRY LAW, C.E. Together with a Series of Tables for Navigation and Nautical Astronomy. By J. R. YOUNG, formerly Professor of Mathematics in Belfast College. New Edition. 12mo, 4s. cloth boards.

Navigation (Practical), with Tables.

PRACTICAL NAVIGATION: consisting of the Sailor's Sea-Book, by JAMES GREENWOOD and W. H. ROSSER; together with the requisite Mathematical and Nautical Tables for the Working of the Problems. By HENRY LAW, C.E., and Professor J. R. YOUNG. Illustrated with numerous Wood Engravings and Coloured Plates. 12mo, 7s. strongly half bound in leather.

WEALE'S RUDIMENTARY SERIES.

The following books in Naval Architecture, etc., are published in the above series.

MASTING, MAST-MAKING, AND RIGGING OF SHIPS. By ROBERT KIPPING, N.A. Fourteenth Edition. 12mo, 2s. 6d. cloth.

SAILS AND SAIL-MAKING. Tenth Edition, enlarged. By ROBERT KIPPING, N.A. Illustrated. 12mo, 3s. cloth boards.

NAVAL ARCHITECTURE. By JAMES PEAKE. Fourth Edition, with Plates and Diagrams. 12mo, 4s. cloth boards.

MARINE ENGINES, AND STEAM VESSELS. By ROBERT MURRAY, C.E. Seventh Edition. 12mo, 3s. 6d. cloth boards.

ARCHITECTURE, BUILDING, ETC.

Construction.
THE SCIENCE of BUILDING: An Elementary Treatise on the Principles of Construction. By E. WYNDHAM TARN, M.A., Architect. With 58 Wood Engravings. Second Edition, revised and enlarged, including an entirely new chapter on the Nature of Lightning, and the Means of Protecting Buildings from its Violence. Crown 8vo, 7s. 6d. cloth.
"A very valuable book, which we strongly recommend to all students."—*Builder.*
"No architectural student should be without this hand-book."—*Architect.*

Civil and Ecclesiastical Building.
A BOOK ON BUILDING, CIVIL AND ECCLESIASTICAL, Including CHURCH RESTORATION. By Sir EDMUND BECKETT, Bart., LL.D., Q.C., F.R.A.S. Author of "Clocks and Watches and Bells," &c. 12mo, 5s. cloth boards.
"A book which is always amusing and nearly always instructive. We are able very cordially to recommend all persons to read it for themselves."—*Times.*
"We commend the book to the thoughtful consideration of all who are interested in the building art."—*Builder.*

Villa Architecture.
A HANDY BOOK of VILLA ARCHITECTURE; being a Series of Designs for Villa Residences in various Styles. With Detailed Specifications and Estimates. By C. WICKES, Architect, Author of "The Spires and Towers of the Mediæval Churches of England," &c. 30 Plates, 4to, half morocco, gilt edges, 1l. 1s.
** Also an Enlarged edition of the above. 61 Plates, with Detailed Specifications, Estimates, &c. 2l. 2s. half morocco.

Useful Text-Book for Architects.
THE ARCHITECT'S GUIDE: Being a Text-book of Useful Information for Architects, Engineers, Surveyors, Contractors, Clerks of Works, &c. By F. ROGERS. Cr. 8vo, 6s. cloth.
"As a text-book of useful information for architects, engineers, surveyors, &c., it would be hard to find a handier or more complete little volume."—*Standard.*

The Young Architect's Book.
HINTS TO YOUNG ARCHITECTS. By G. WIGHTWICK. New Edition. By G. H. GUILLAUME. 12mo, cloth, 4s.
"Will be found an acquisition to pupils, and a copy ought to be considered as necessary a purchase as a box of instruments."—*Architect.*

Drawing for Builders and Students.
PRACTICAL RULES ON DRAWING for the OPERATIVE BUILDER and YOUNG STUDENT in ARCHITECTURE. By GEORGE PYNE. With 14 Plates, 4to, 7s. 6d. boards.

Boiler and Factory Chimneys.
BOILER AND FACTORY CHIMNEYS; their Draught-power and Stability, with a chapter on Lightning Conductors. By ROBERT WILSON, C.E. Crown 8vo, 3s. 6d. cloth.

Taylor and Cresy's Rome.

THE ARCHITECTURAL ANTIQUITIES OF ROME. By the late G. L. TAYLOR, Esq., F.S.A., and EDWARD CRESY, Esq. New Edition, Edited by the Rev. ALEXANDER TAYLOR, M.A. (son of the late G. L. Taylor, Esq.) This is the only book which gives on a large scale, and with the precision of architectural measurement, the principal Monuments of Ancient Rome in plan, elevation, and detail. Large folio, with 130 Plates, half-bound, 3*l*. 3*s*.

⁎ Originally published in two volumes, folio, at 18*l*. 18*s*.

Vitruvius' Architecture.

THE ARCHITECTURE OF MARCUS VITRUVIUS POLLIO. Translated by JOSEPH GWILT, F.S.A., F.R.A.S. Numerous Plates. 12mo, cloth limp, 5*s*.

Ancient Architecture.

RUDIMENTARY ARCHITECTURE (ANCIENT); comprising VITRUVIUS, translated by JOSEPH GWILT, F.S.A., &c., with 23 fine plates; and GRECIAN ARCHITECTURE. By the EARL of ABERDEEN; 12mo, 6*s*., half-bound.

⁎ *The only edition of VITRUVIUS procurable at a moderate price.*

Modern Architecture.

RUDIMENTARY ARCHITECTURE (MODERN); comprising THE ORDERS OF ARCHITECTURE. By W. H. LEEDS, Esq.; The STYLES of ARCHITECTURE of VARIOUS COUNTRIES. By T. TALBOT BURY; and The PRINCIPLES of DESIGN in ARCHITECTURE. By E. L. GARBETT. Numerous illustrations, 12mo, 6*s*. half-bound.

Civil Architecture.

A TREATISE on THE DECORATIVE PART of CIVIL ARCHITECTURE. By Sir WILLIAM CHAMBERS, F.R.S. With Illustrations, Notes, and an Examination of Grecian Architecture. By JOSEPH GWILT, F.S.A. Revised and edited by W. H. LEEDS. 66 Plates, 4to, 21*s*. cloth.

House Painting.

HOUSE PAINTING, GRAINING, MARBLING, AND SIGN WRITING: a Practical Manual of. With 9 Coloured Plates of Woods and Marbles, and nearly 150 Wood Engravings. By ELLIS A. DAVIDSON. Third Edition, Revised. 12mo, 6*s*. cloth.

Plumbing.

PLUMBING; a Text-book to the Practice of the Art or Craft of the Plumber. With chapters upon House-drainage, embodying the latest Improvements. By W. P. BUCHAN, Sanitary Engineer. Fourth Edition, Revised and much enlarged, with 300 illustrations, 12mo. 4*s*. cloth. [*Just published.*

Joints used in Building, Engineering, &c.

THE JOINTS MADE AND USED BY BUILDERS in the construction of various kinds of Engineering and Architectural works, with especial reference to those wrought by artificers in erecting and finishing Habitable Structures. By W. J. CHRISTY, Architect. With 160 Illustrations. 12mo, 3*s*. 6*d*. cloth boards.

[*Just published.*

Handbook of Specifications.

THE HANDBOOK OF SPECIFICATIONS; or, Practical Guide to the Architect, Engineer, Surveyor, and Builder, in drawing up Specifications and Contracts for Works and Constructions. Illustrated by Precedents of Buildings actually executed by eminent Architects and Engineers. By Professor THOMAS L. DONALDSON, M.I.B.A. New Edition, in One large volume, 8vo, with upwards of 1000 pages of text, and 33 Plates, cloth, 1*l*. 11*s*. 6*d*.

"In this work forty-four specifications of executed works are given. . . . Donaldson's Handbook of Specifications must be bought by all architects."—*Builder.*

Specifications for Practical Architecture.

SPECIFICATIONS FOR PRACTICAL ARCHITECTURE: A Guide to the Architect, Engineer, Surveyor, and Builder; with an Essay on the Structure and Science of Modern Buildings. By FREDERICK ROGERS, Architect. 8vo, 15*s*. cloth.

** A volume of specifications of a practical character being greatly required, and the old standard work of Alfred Bartholomew being out of print, the author, on the basis of that work, has produced the above.—*Extract from Preface.*

Designing, Measuring, and Valuing.

THE STUDENT'S GUIDE to the PRACTICE of MEASURING and VALUING ARTIFICERS' WORKS; containing Directions for taking Dimensions, Abstracting the same, and bringing the Quantities into Bill, with Tables of Constants, and copious Memoranda for the Valuation of Labour and Materials in the respective Trades of Bricklayer and Slater, Carpenter and Joiner, Painter and Glazier, Paperhanger, &c. With 43 Plates and Woodcuts. Originally edited by EDWARD DOBSON, Architect. New Edition, re-written, with Additions on Mensuration and Construction, and useful Tables for facilitating Calculations and Measurements. By E. WYNDHAM TARN, M.A., 8vo, 10*s*. 6*d*. cloth.

"Well fulfils the promise of its title-page. Mr. Tarn's additions and revisions have much increased the usefulness of the work."—*Engineering.*

Beaton's Pocket Estimator.

THE POCKET ESTIMATOR FOR THE BUILDING TRADES, being an easy method of estimating the various parts of a Building collectively, more especially applied to Carpenters' and Joiners' work. By A. C. BEATON. Second Edition. Waistcoat-pocket size. 1*s*. 6*d*.

Beaton's Builders' and Surveyors' Technical Guide.

THE POCKET TECHNICAL GUIDE AND MEASURER FOR BUILDERS AND SURVEYORS: containing a Complete Explanation of the Terms used in Building Construction, Memoranda for Reference, Technical Directions for Measuring Work in all the Building Trades, &c. By A. C. BEATON. 1*s*. 6*d*.

The House-Owner's Estimator.

THE HOUSE-OWNER'S ESTIMATOR; or, What will it Cost to Build, Alter, or Repair? A Price-Book for Unprofessional People, Architectural Surveyors, Builders, &c. By the late JAMES D. SIMON. Edited by F. T. W. MILLER, A.R.I.B.A. Third Edition, Revised. Crown 8vo, 3*s*. 6*d*., cloth.

"In two years it will repay its cost a hundred times over."—*Field.*

Cement.

PORTLAND CEMENT FOR USERS. By HENRY FAIJA, A.M. Inst. C.E., with Illustrations. Crown 8vo. 3s. 6d. cloth.
"A useful compendium of results for the builder and architect."—*Building News.*

Builder's and Contractor's Price Book.

LOCKWOOD & CO.'S BUILDER'S AND CONTRACTOR'S PRICE BOOK, containing the latest prices of all kinds of Builders' Materials and Labour, and of all Trades connected with Building, &c. Revised by F. T. W. MILLER, A.R.I.B.A. Half-bound, 4s.

CARPENTRY, TIMBER, ETC.

Tredgold's Carpentry, new and cheaper Edition.

THE ELEMENTARY PRINCIPLES OF CARPENTRY: a Treatise on the Pressure and Equilibrium of Timber Framing, the Resistance of Timber, and the Construction of Floors, Arches, Bridges, Roofs, Uniting Iron and Stone with Timber, &c. To which is added an Essay on the Nature and Properties of Timber, &c., with Descriptions of the Kinds of Wood used in Building; also numerous Tables of the Scantlings of Timber for different purposes, the Specific Gravities of Materials, &c. By THOMAS TREDGOLD, C.E. Edited by PETER BARLOW, F.R.S. Fifth Edition, corrected and enlarged. With 64 Plates, Portrait of the Author, and Woodcuts. 4to, published at 2l. 2s., reduced to 1l. 5s. cloth.
"Ought to be in every architect's and every builder's library, and those who do not already possess it ought to avail themselves of the new issue."—*Builder.*
"A work whose monumental excellence must commend it wherever skilful carpentry is concerned. The Author's principles are rather confirmed than impaired by time. The additional plates are of great intrinsic value."—*Building News.*

Grandy's Timber Tables.

THE TIMBER IMPORTER'S, TIMBER MERCHANT'S, & BUILDER'S STANDARD GUIDE. By R. E. GRANDY. 2nd Edition. Carefully revised and corrected. 12mo, 3s. 6d. cloth.
"Everything it pretends to be: built up gradually, it leads one from a forest to a treenail, and throws in, as a makeweight, a host of material concerning bricks, columns, cisterns, &c.—all that the class to whom it appeals requires."—*English Mechanic.*

Timber Freight Book.

THE TIMBER IMPORTERS' AND SHIPOWNERS' FREIGHT BOOK: Being a Comprehensive Series of Tables for the Use of Timber Importers, Captains of Ships, Shipbrokers, Builders, and Others. By W. RICHARDSON. Crown 8vo, 6s.

Tables for Packing-Case Makers.

PACKING-CASE TABLES; showing the number of Superficial Feet in Boxes or Packing-Cases, from six inches square and upwards. By W. RICHARDSON. Oblong 4to, 3s. 6d. cloth.
"Invaluable labour-saving tables."—*Ironmonger.*

Carriage Building, &c.

COACH BUILDING: A Practical Treatise, Historical and Descriptive, containing full information of the various Trades and Processes involved, with Hints on the proper keeping of Carriages, &c. 57 Illustrations. By JAMES W. BURGESS. 12mo, 3s. cloth.

Horton's Measurer.

THE COMPLETE MEASURER; setting forth the Measurement of Boards, Glass, &c.; Unequal-sided, Square-sided, Octagonal-sided, Round Timber and Stone, and Standing Timber. With just allowances for the bark in the respective species of trees, and proper deductions for the waste in hewing the trees, &c.; also a Table showing the solidity of hewn or eight-sided timber, or of any octagonal-sided column. By RICHARD HORTON. Fourth Edition, with considerable and valuable additions, 12mo, strongly bound in leather, 5s.

Horton's Underwood and Woodland Tables.

TABLES FOR PLANTING AND VALUING UNDERWOOD AND WOODLAND; also Lineal, Superficial, Cubical, and Decimal Tables, &c. By R. HORTON. 12mo, 2s. leather.

Nicholson's Carpenter's Guide.

THE CARPENTER'S NEW GUIDE; or, BOOK of LINES for CARPENTERS: comprising all the Elementary Principles essential for acquiring a knowledge of Carpentry. Founded on the late PETER NICHOLSON'S standard work. A new Edition, revised by ARTHUR ASHPITEL, F.S.A., together with Practical Rules on Drawing, by GEORGE PYNE. With 74 Plates, 4to, 1l. 1s. cloth.

Dowsing's Timber Merchant's Companion.

THE TIMBER MERCHANT'S AND BUILDER'S COMPANION; containing New and Copious Tables of the Reduced Weight and Measurement of Deals and Battens, of all sizes, from One to a Thousand Pieces, also the relative Price that each size bears per Lineal Foot to any given Price per Petersburgh Standard Hundred, &c., &c. Also a variety of other valuable information. By WILLIAM DOWSING, Timber Merchant. Third Edition, Revised. Crown 8vo, 3s. cloth.

"Everything is as concise and clear as it can possibly be made. There can be no doubt that every timber merchant and builder ought to possess it."—*Hull Advertiser.*

Practical Timber Merchant.

THE PRACTICAL TIMBER MERCHANT, being a Guide for the use of Building Contractors, Surveyors, Builders, &c., comprising useful Tables for all purposes connected with the Timber Trade, Essay on the Strength of Timber, Remarks on the Growth of Timber, &c. By W. RICHARDSON. Fcap. 8vo, 3s. 6d. cl.

Woodworking Machinery.

WOODWORKING MACHINERY; its Rise, Progress, and Construction. With Hints on the Management of Saw Mills and the Economical Conversion of Timber. Illustrated with Examples of Recent Designs by leading English, French, and American Engineers. By M. POWIS BALE, M.I.M.E. Large crown 8vo, 12s. 6d. cloth.

"Mr. Bale is evidently an expert on the subject, and he has collected so much information that his book is all-sufficient for builders and others engaged in the conversion of timber."—*Architect.*

"The most comprehensive compendium of wood-working machinery we have seen. The author is a thorough master of his subject."—*Building News.*

"It should be in the office of every wood-working factory."—*English Mechanic.*

C

MECHANICS, ETC.

Turning.
LATHE-WORK: a Practical Treatise on the Tools, Appliances, and Processes employed in the Art of Turning. By PAUL N. HASLUCK. With Illustrations drawn by the Author. Crown 8vo, 5s.
"Evidently written from personal experience, and gives a large amount of just that sort of information which beginners at the lathe require."—*Builder.*

Mechanic's Workshop Companion.
THE OPERATIVE MECHANIC'S WORKSHOP COMPANION, and THE SCIENTIFIC GENTLEMAN'S PRACTICAL ASSISTANT. By W. TEMPLETON. 13th Edit., with Mechanical Tables for Operative Smiths, Millwrights, Engineers, &c.; and an Extensive Table of Powers and Roots, 12mo, 5s. bound.
"Admirably adapted to the wants of a very large class. It has met with great success in the engineering workshop, as we can testify; and there are a great many men who, in a great measure, owe their rise in life to this little work."—*Building News.*

Engineer's and Machinist's Assistant.
THE ENGINEER'S, MILLWRIGHT'S, and MACHINIST'S PRACTICAL ASSISTANT; comprising a Collection of Useful Tables, Rules, and Data. By WM. TEMPLETON. 18mo, 2s. 6d.

Smith's Tables for Mechanics, &c.
TABLES, MEMORANDA, and CALCULATED RESULTS, FOR MECHANICS, ENGINEERS, ARCHITECTS, BUILDERS, &c. Selected and Arranged by FRANCIS SMITH. Waistcoat-pocket size, 1s. 6d., limp leather. [*Just published.*

Boiler Making.
THE BOILER-MAKER'S READY RECKONER. With Examples of Practical Geometry and Templating, for the use of Platers, Smiths, and Riveters. By JOHN COURTNEY, Edited by D. K. CLARK, M.I.C.E. 12mo, 9s. half-bd. [*Just published.*

Superficial Measurement.
THE TRADESMAN'S GUIDE TO SUPERFICIAL MEASUREMENT. Tables calculated from 1 to 200 inches in length, by 1 to 108 inches in breadth. By J. HAWKINGS. Fcp. 3s. 6d. cl.

The High-Pressure Steam Engine.
THE HIGH-PRESSURE STEAM ENGINE. By Dr. ERNST ALBAN. Translated from the German, with Notes, by Dr. POLE, F.R.S. Plates, 8vo, 16s. 6d. cloth.

Steam Boilers.
A TREATISE ON STEAM BOILERS: their Strength, Construction, and Economical Working. By R. WILSON, C.E. Fifth Edition. 12mo, 6s. cloth.
"The best work on boilers which has come under our notice."—*Engineering.*
"The best treatise that has ever been published on steam boilers."—*Engineer.*

Mechanics.
THE HANDBOOK OF MECHANICS. By DIONYSIUS LARDNER, D.C.L. New Edition, Edited and considerably Enlarged, by BENJAMIN LOEWY, F.R.A.S., &c., post 8vo, 6s. cloth.

MATHEMATICS, TABLES, ETC.

Metrical Units and Systems, &c.
MODERN METROLOGY: A Manual of the Metrical Units and Systems of the present Century. With an Appendix containing a proposed English System. By LOWIS D'A. JACKSON, A.-M. Inst. C.E., Author of "Aid to Survey Practice," &c. Large Crown 8vo, 12s. 6d. cloth. [*Just published.*

Gregory's Practical Mathematics.
MATHEMATICS for PRACTICAL MEN; being a Commonplace Book of Pure and Mixed Mathematics. Designed chiefly for the Use of Civil Engineers, Architects, and Surveyors. Part I. PURE MATHEMATICS—comprising Arithmetic, Algebra, Geometry, Mensuration, Trigonometry, Conic Sections, Properties of Curves. Part II. MIXED MATHEMATICS—comprising Mechanics in general, Statics, Dynamics, Hydrostatics, Hydrodynamics, Pneumatics, Mechanical Agents, Strength of Materials, &c. By OLINTHUS GREGORY, LL.D., F.R.A.S. Enlarged by H. LAW, C.E. 4th Edition, revised by Prof. J. R. YOUNG. With 13 Plates. 8vo, 1l. 1s. cloth.

Mathematics as applied to the Constructive Arts.
A TREATISE ON MATHEMATICS AS APPLIED TO THE CONSTRUCTIVE ARTS. Illustrating the various processes of Mathematical Investigation by means of Arithmetical and simple Algebraical Equations and Practical Examples, &c. By FRANCIS CAMPIN, C.E. 12mo, 3s. 6d. cloth. [*Just published.*

Geometry for the Architect, Engineer, &c.
PRACTICAL GEOMETRY, for the Architect, Engineer, and Mechanic. By E. W. TARN, M.A. With Appendices on Diagrams of Strains and Isometrical projection. Demy 8vo, 9s. cloth.

The Metric System.
A SERIES OF METRIC TABLES, in which the British Standard Measures and Weights are compared with those of the Metric System at present in use on the Continent. By C. H. DOWLING, C.E. 2nd Edit., revised and enlarged. 8vo, 10s. 6d. cl.

Inwood's Tables, greatly enlarged and improved.
TABLES FOR THE PURCHASING of ESTATES, Freehold, Copyhold, or Leasehold; Annuities, Advowsons, &c., and for the Renewing of Leases; also for Valuing Reversionary Estates, Deferred Annuities, &c. By WILLIAM INWOOD. 21st edition, with Tables of Logarithms for the more Difficult Computations of the Interest of Money, &c. By M. FÉDOR THOMAN. 12mo. 8s. cloth.
"Those interested in the purchase and sale of estates, and in the adjustment of compensation cases, as well as in transactions in annuities, life insurances, &c., will find the present edition of eminent service."—*Engineering.*

Weights, Measures, Moneys, &c.
MEASURES, WEIGHTS, and MONEYS of all NATIONS, and an Analysis of the Christian, Hebrew, and Mahometan Calendars. Entirely New Edition, Revised and Enlarged. By W. S. B. WOOLHOUSE, F.R.A.S. 12mo, 2s. 6d. cloth boards.

Compound Interest and Annuities.

THEORY of COMPOUND INTEREST and ANNUITIES; with Tables of Logarithms for the more Difficult Computations of Interest, Discount, Annuities, &c., in all their Applications and Uses for Mercantile and State Purposes. By Fédor Thoman, of the Société Crédit Mobilier, Paris. 3rd Edit., 12mo, 4s. 6d. cl.

Iron and Metal Trades' Calculator.

THE IRON AND METAL TRADES' COMPANION: Being a Calculator containing a Series of Tables upon a new and comprehensive plan for expeditiously ascertaining the value of any goods bought or sold by weight, from 1s. per cwt. to 112s. per cwt., and from one farthing per lb. to 1s. per lb. Each Table extends from one lb. to 100 tons. By T. Downie. 396 pp., 9s., leather.
"Will supply a want, for nothing like it before existed."—*Building News.*

Iron and Steel.

IRON AND STEEL: a Work for the Forge, Foundry, Factory, and Office. Containing Information for Ironmasters and their Stocktakers; Managers of Bar, Rail, Plate, and Sheet Rolling Mills; Iron and Metal Founders; Iron Ship and Bridge Builders; Mechanical, Mining, and Consulting Engineers; Architects, Builders, &c. By Charles Hoare, Author of 'The Slide Rule,' &c. Eighth Edition. Oblong 32mo, 6s., leather.
"For comprehensiveness the book has not its equal."—*Iron.*

Comprehensive Weight Calculator.

THE WEIGHT CALCULATOR, being a Series of Tables upon a New and Comprehensive Plan, exhibiting at one Reference the exact Value of any Weight from 1 lb. to 15 tons, at 300 Progressive Rates, from 1 Penny to 168 Shillings per cwt., and containing 186,000 Direct Answers, which, with their Combinations, consisting of a single addition (mostly to be performed at sight), will afford an aggregate of 10,266,000 Answers; the whole being calculated and designed to ensure Correctness and promote Despatch. By Henry Harben, Accountant, Sheffield. New Edition. Royal 8vo, 1l. 5s., strongly half-bound.

Comprehensive Discount Guide.

THE DISCOUNT GUIDE: comprising several Series of Tables for the use of Merchants, Manufacturers, Ironmongers, and others, by which may be ascertained the exact profit arising from any mode of using Discounts, either in the Purchase or Sale of Goods, and the method of either Altering a Rate of Discount, or Advancing a Price, so as to produce, by one operation, a sum that will realise any required profit after allowing one or more Discounts: to which are added Tables of Profit or Advance from 1¼ to 90 per cent., Tables of Discount from 1¼ to 98⅞ per cent., and Tables of Commission, &c., from ⅛ to 10 per cent. By Henry Harben, Accountant. New Edition. Demy 8vo, 1l. 5s., half-bound.

Mathematical Instruments.

MATHEMATICAL INSTRUMENTS: Their Construction, Adjustment, Testing, and Use; comprising Drawing, Measuring, Optical, Surveying, and Astronomical Instruments. By J. F. Heather, M.A. Enlarged Edition. 12mo, 5s. cloth.

SCIENCE AND ART.

Gold and Gold-Working.
THE GOLDSMITH'S HANDBOOK: containing full instructions for the Alloying and Working of Gold. Including the Art of Alloying, Melting, Reducing, Colouring, Collecting and Refining. Chemical and Physical Properties of Gold, with a new System of Mixing its Alloys; Solders, Enamels, &c. By GEORGE E. GEE. Second Edition, enlarged. 12mo, 3s. 6d. cloth.
" The best work yet printed on its subject for a reasonable price."—*Jeweller.*
" Essentially a practical manual, well adapted to the wants of amateurs and apprentices, containing trustworthy information that only a practical man can supply."—*English Mechanic.*

Silver and Silver Working.
THE SILVERSMITH'S HANDBOOK, containing full Instructions for the Alloying and Working of Silver. Including the different Modes of Refining and Melting the Metal, its Solders, the Preparation of Imitation Alloys, &c. By G. E. GEE. 12mo, 3s. 6d.
" The chief merit of the work is its practical character. The workers in the trade will speedily discover its merits when they sit down to study it."—*English Mechanic*

Hall-Marking of Jewellery.
THE HALL-MARKING OF JEWELLERY PRACTICALLY CONSIDERED, comprising an account of all the different Assay Towns of the United Kingdom; with the Stamps at present employed; also the Laws relating to the Standards and Hall-Marks at the various Assay Offices; and a variety of Practical Suggestions concerning the Mixing of Standard Alloys, &c. By GEORGE E. GEE. Crown 8vo, 5s. cloth. [*Just published.*

Electro-Plating, &c.
ELECTROPLATING: A Practical Handbook. By J. W. URQUHART, C.E. Crown 8vo, 5s. cloth.
" Any ordinarily intelligent person may become an adept in electro-deposition with a very little science indeed, and this is the book to show the way."—*Builder.*

Electrotyping, &c.
ELECTROTYPING: A Practical Manual on the Reproduction and Multiplication of Printing Surfaces and Works of Art by the Electro-deposition of Metals. By J. W. URQUHART, C.E. Crown 8vo, 5s. cloth.
"A guide to beginners and those who practise the old and imperfect methods."—*Iron.*

Electro-Plating.
ELECTRO-METALLURGY PRACTICALLY TREATED. By ALEXANDER WATT, F.R.S.S.A. Including the Electro-Deposition of Copper, Silver, Gold, Brass and Bronze, Platinum, Lead, Nickel, Tin, Zinc, Alloys of Metals, Practical Notes, &c., &c. Eighth Edition, Revised and Enlarged, including the most recent Processes. 12mo, 3s. 6d. cloth. [*Just published.*
" From this book both amateur and artisan may learn everything necessary for the successful prosecution of electroplating."—*Iron.*
" A practical treatise for the use of those who desire to work in the art of electrodeposition as a business."—*English Mechanic.*

WORKS IN SCIENCE AND ART, ETC.,

Dentistry.

MECHANICAL DENTISTRY. A Practical Treatise on the Construction of the various kinds of Artificial Dentures. Comprising also Useful Formulæ, Tables, and Receipts for Gold Plate, Clasps, Solders, etc., etc. By CHARLES HUNTER. Second Edition, Revised; including a new chapter on the use of Celluloid. With over 100 Engravings. Cr. 8vo, 7s. 6d. cl. [*Just published*.
"An authoritative treatise, which we can strongly recommend to all students."—*Dublin Journal of Medical Science.*
"The best book on the subject with which we are acquainted."—*Medical Press.*

Electricity.

A MANUAL of ELECTRICITY; including Galvanism, Magnetism, Diamagnetism, Electro-Dynamics, Magneto-Electricity, and the Electric Telegraph. By HENRY M. NOAD, Ph.D., F.C.S. Fourth Edition, with 500 Woodcuts. 8vo, 1l. 4s. cloth.
"The accounts given of electricity and galvanism are not only complete in a scientific sense, but, which is a rarer thing, are popular and interesting."—*Lancet.*

Text-Book of Electricity.

THE STUDENT'S TEXT-BOOK OF ELECTRICITY. By HENRY M. NOAD, Ph.D., F.R.S., &c. New Edition, Revised. With an Introduction and Additional Chapters by W. H. PREECE, M.I.C.E., Vice-President of the Society of Telegraph Engineers, &c. With 470 Illustrations. Crown 8vo, 12s. 6d. cloth.
"A reflex of the existing state of Electrical Science adapted for students."— W. H. Preece, Esq., vide "Introduction."
"We can recommend Dr. Noad's book for clear style, great range of subject, a good index, and a plethora of woodcuts. Such collections as the present are indispensable."—*Athenæum.*
"An admirable text-book for every student—beginner or advanced—of electricity."
—*Engineering.*
"Recommended to students as one of the best text-books on the subject that they can have. Mr. Preece appears to have introduced all the newest inventions in the shape of telegraphic, telephonic, and electric-lighting apparatus."—*English Mechanic.*
"Under the editorial hand of Mr. Preece the late Dr. Noad's text-book of electricity has grown into an admirable handbook."—*Westminster Review.*

Electric Lighting.

ELECTRIC LIGHT: Its Production and Use, embodying plain Directions for the Working of Galvanic Batteries, Electric Lamps, and Dynamo-Electric Machines. By J. W. URQUHART, C.E., Author of "Electroplating." Edited by F. C. WEBB, M.I.C.E., M.S.T.E. With 94 Illustrations. Crown 8vo, 7s. 6d. cloth.
"The book is by far the best that we have yet met with on the subject."—*Athenæum.*
"An important addition to the literature of the electric light. Students of the subject should not fail to read it."—*Colliery Guardian.*

Lightning, &c.

THE ACTION of LIGHTNING, and the MEANS of DEFENDING LIFE AND PROPERTY FROM ITS EFFECTS. By ARTHUR PARNELL, Major in the Corps of Royal Engineers. 12mo, 7s. 6d. cloth. [*Just published.*
"Major Parnell has written an original work on a scientific subject of unusual interest; and he has prefaced his arguments by a patient and almost exhaustive citation of the best writers on the subject in the English language."—*Athenæum.*
"The work comprises all that is actually known on the subject."—*Land.*
"Major Parnell's measures are based on the results of experience. A valuable *repertoire* of facts and principles arranged in a scientific form."—*Building News.*

The Alkali Trade—Sulphuric Acid, &c.

A MANUAL OF THE ALKALI TRADE, including the Manufacture of Sulphuric Acid, Sulphate of Soda, and Bleaching Powder. By JOHN LOMAS, Alkali Manufacturer, Newcastle-upon-Tyne and London. With 232 Illustrations and Working Drawings, and containing 386 pages of text. Super-royal 8vo, 2*l*. 12*s*. 6*d*. cloth.

This work provides (1) a Complete Handbook for intending Alkali and Sulphuric Acid Manufacturers, and for those already in the field who desire to improve their plant, or to become practically acquainted with the latest processes and developments of the trade; (2) a Handy Volume which Manufacturers can put into the hands of their Managers and Foremen as a useful guide in their daily rounds of duty.

SYNOPSIS OF CONTENTS.

Chap. I. Choice of Site and General Plan of Works—II. Sulphuric Acid—III. Recovery of the Nitrogen Compounds, and Treatment of Small Pyrites—IV. The Salt Cake Process—V. Legislation upon the Noxious Vapours Question—VI. The Hargreaves' and Jones' Processes—VII. The Balling Process—VIII. Lixiviation and Salting Down—IX. Carbonating or Finishing—X. Soda Crystals — XI. Refined Alkali — XII. Caustic Soda — XIII. Bi-carbonate of Soda — XIV. Bleaching Powder — XV. Utilisation of Tank Waste—XVI. General Remarks—Four Appendices, treating of Yields, Sulphuric Acid Calculations, Anemometers, and Foreign Legislation upon the Noxious Vapours Question.

"The author has given the fullest, most practical, and, to all concerned in the alkali trade, most valuable mass of information that, to our knowledge, has been published in any language."—*Engineer.*

"This book is written by a manufacturer for manufacturers. The working details of the most approved forms of apparatus are given, and these are accompanied by no less than 232 wood engravings, all of which may be used for the purposes of construction. Every step in the manufacture is very fully described in this manual, and each improvement explained. Everything which tends to introduce economy into the technical details of this trade receives the fullest attention. The book has been produced with great completeness."—*Athenæum.*

"The author is not one of those clever compilers who, on short notice, will 'read up' any conceivable subject, but a practical man in the best sense of the word. We find here not merely a sound and luminous explanation of the chemical principles of the trade, but a notice of numerous matters which have a most important bearing on the successful conduct of alkali works, but which are generally overlooked by even the most experienced technological authors. This most valuable book, which we trust will be generally appreciated, we must pronounce a credit alike to its author and to the enterprising firm who have undertaken its publication."—*Chemical Review.*

Chemical Analysis.

THE COMMERCIAL HANDBOOK of CHEMICAL ANALYSIS; or Practical Instructions for the determination of the Intrinsic or Commercial Value of Substances used in Manufactures, in Trades, and in the Arts. By A. NORMANDY, Author of "Practical Introduction to Rose's Chemistry," and Editor of Rose's "Treatise on Chemical Analysis." *New Edition.* Enlarged, and to a great extent re-written, by HENRY M. NOAD, Ph. D., F.R.S. With numerous Illustrations. Cr. 8vo, 12*s*. 6*d*. cloth.

"We recommend this book to the careful perusal of every one; it may be truly affirmed to be of universal interest, and we strongly recommend it to our readers as a guide, alike indispensable to the housewife as to the pharmaceutical practitioner."—*Medical Times.*
"Essential to the analysts appointed under the new Act. The most recent results are given, and the work is well edited and carefully written."—*Nature.*

Dr. Lardner's Museum of Science and Art.

THE MUSEUM OF SCIENCE AND ART. Edited by DIONYSIUS LARDNER, D.C.L., formerly Professor of Natural Philosophy and Astronomy in University College, London. With upwards of 1200 Engravings on Wood. In 6 Double Volumes. Price £1 1s., in a new and elegant cloth binding, or handsomely bound in half morocco, 31s. 6d.

OPINIONS OF THE PRESS.

" This series, besides affording popular but sound instruction on scientific subjects, with which the humblest man in the country ought to be acquainted, also undertakes that teaching of 'common things' which every well-wisher of his kind is anxious to promote. Many thousand copies of this serviceable publication have been printed, in the belief and hope that the desire for instruction and improvement widely prevails; and we have no fear that such enlightened faith will meet with disappointment."—*Times.*

" A cheap and interesting publication, alike informing and attractive. The papers combine subjects of importance and great scientific knowledge, considerable inductive powers, and a popular style of treatment."—*Spectator.*

" The 'Museum of Science and Art' is the most valuable contribution that has ever been made to the Scientific Instruction of every class of society."—*Sir David Brewster in the North British Review.*

" Whether we consider the liberality and beauty of the illustrations, the charm of the writing, or the durable interest of the matter, we must express our belief that there is hardly to be found among the new books, one that would be welcomed by people of so many ages and classes as a valuable present."—*Examiner.*

** *Separate books formed from the above, suitable for Workmen's Libraries, Science Classes, &c.*

COMMON THINGS EXPLAINED. Containing Air, Earth, Fire, Water, Time, Man, the Eye, Locomotion, Colour, Clocks and Watches, &c. 233 Illustrations, cloth gilt, 5s.

THE MICROSCOPE. Containing Optical Images, Magnifying Glasses, Origin and Description of the Microscope, Microscopic Objects, the Solar Microscope, Microscopic Drawing and Engraving, &c. 147 Illustrations, cloth gilt, 2s.

POPULAR GEOLOGY. Containing Earthquakes and Volcanoes, the Crust of the Earth, etc. 201 Illustrations, cloth gilt, 2s. 6d.

POPULAR PHYSICS. Containing Magnitude and Minuteness, the Atmosphere, Meteoric Stones, Popular Fallacies, Weather Prognostics, the Thermometer, the Barometer, Sound, &c. 85 Illustrations, cloth gilt, 2s. 6d.

STEAM AND ITS USES. Including the Steam Engine, the Locomotive, and Steam Navigation. 89 Illustrations, cloth gilt, 2s.

POPULAR ASTRONOMY. Containing How to Observe the Heavens. The Earth, Sun, Moon, Planets. Light, Comets, Eclipses, Astronomical Influences, &c. 182 Illustrations, 4s. 6d.

THE BEE AND WHITE ANTS: Their Manners and Habits. With Illustrations of Animal Instinct and Intelligence. 135 Illustrations, cloth gilt, 2s.

THE ELECTRIC TELEGRAPH POPULARISED. To render intelligible to all who can Read, irrespective of any previous Scientific Acquirements, the various forms of Telegraphy in Actual Operation. 100 Illustrations, cloth gilt, 1s. 6d.

Dr. Lardner's Handbooks of Natural Philosophy.

*** *The following five volumes, though each is Complete in itself, and to be purchased separately, form* A COMPLETE COURSE OF NATURAL PHILOSOPHY, *and are intended for the general reader who desires to attain accurate knowledge of the various departments of Physical Science, without pursuing them according to the more profound methods of mathematical investigation. The style is studiously popular. It has been the author's aim to supply Manuals such as are required by the Student, the Engineer, the Artisan, and the superior classes in Schools.*

THE HANDBOOK OF MECHANICS. Enlarged and almost rewritten by BENJAMIN LOEWY, F.R.A.S. With 378 Illustrations. Post 8vo, 6s. cloth.

"The perspicuity of the original has been retained, and chapters which had become obsolete, have been replaced by others of more modern character. The explanations throughout are studiously popular, and care has been taken to show the application of the various branches of physics to the industrial arts, and to the practical business of life."—*Mining Journal.*

THE HANDBOOK of HYDROSTATICS and PNEUMATICS. New Edition, Revised and Enlarged by BENJAMIN LOEWY, F.R.A.S. With 236 Illustrations. Post 8vo, 5s. cloth.

" For those ' who desire to attain an accurate knowledge of physical science without the profound methods of mathematical investigation,' this work is not merely intended, but well adapted."—*Chemical News.*

THE HANDBOOK OF HEAT. Edited and almost entirely Rewritten by BENJAMIN LOEWY, F.R.A.S., etc. 117 Illustrations. Post 8vo, 6s. cloth.

"The style is always clear and precise, and conveys instruction without leaving any cloudiness or lurking doubts behind."—*Engineering.*

THE HANDBOOK OF OPTICS. New Edition. Edited by T. OLVER HARDING, B.A. 298 Illustrations. Post 8vo, 5s. cloth.

" Written by one of the ablest English scientific writers, beautifully and elaborately illustrated."—*Mechanics' Magazine.*

THE HANDBOOK OF ELECTRICITY, MAGNETISM, and ACOUSTICS. New Edition. Edited by GEO. CAREY FOSTER, B.A., F.C.S. With 400 Illustrations. Post 8vo, 5s. cloth.

" The book could not have been entrusted to any one better calculated to preserve the terse and lucid style of Lardner, while correcting his errors and bringing up his work to the present state of scientific knowledge."—*Popular Science Review.*

Dr. Lardner's Handbook of Astronomy.

THE HANDBOOK OF ASTRONOMY. Forming a Companion to the "Handbooks of Natural Philosophy." By DIONYSIUS LARDNER, D.C.L. Fourth Edition. Revised and Edited by EDWIN DUNKIN, F.R.S., Royal Observatory, Greenwich. With 38 Plates and upwards of 100 Woodcuts. In 1 vol., small 8vo, 550 pages, 9s. 6d., cloth.

" Probably no other book contains the same amount of information in so compendious and well-arranged a form—certainly none at the price at which this is offered to the public."—*Athenæum.*

" We can do no other than pronounce this work a most valuable manual of astronomy, and we strongly recommend it to all who wish to acquire a general—but at the same time correct—acquaintance with this sublime science."—*Quarterly Journal of Science.*

Dr. Lardner's Handbook of Animal Physics.

THE HANDBOOK OF ANIMAL PHYSICS. By DR. LARDNER. With 520 Illustrations. New edition, small 8vo, cloth, 732 pages, 7s. 6d.

" We have no hesitation in cordially recommending it."—*Educational Times.*

Dr. Lardner's School Handbooks.
NATURAL PHILOSOPHY FOR SCHOOLS. By Dr. Lardner. 328 Illustrations. Sixth Edition. 1 vol. 3s. 6d. cloth.
"Conveys, in clear and precise terms, general notions of all the principal divisions of Physical Science."—*British Quarterly Review.*
ANIMAL PHYSIOLOGY FOR SCHOOLS. By Dr. Lardner. With 190 Illustrations. Second Edition. 1 vol. 3s. 6d. cloth.
"Clearly written, well arranged, and excellently illustrated."—*Gardeners' Chronicle.*

Dr. Lardner's Electric Telegraph.
THE ELECTRIC TELEGRAPH. By Dr. Lardner. New Edition. Revised and Re-written, by E. B. Bright, F.R.A.S, 140 Illustrations. Small 8vo, 2s. 6d. cloth.
"One of the most readable books extant on the Electric Telegraph."—*Eng. Mechanic.*

Mollusca.
A MANUAL OF THE MOLLUSCA; being a Treatise on Recent and Fossil Shells. By Dr. S. P. Woodward, A.L.S. With Appendix by Ralph Tate, A.L.S., F.G.S. With numerous Plates and 300 Woodcuts. 3rd Edition. Cr. 8vo, 7s. 6d. cloth.

Geology and Genesis.
THE TWIN RECORDS OF CREATION; or, Geology and Genesis, their Perfect Harmony and Wonderful Concord. By George W. Victor le Vaux. Fcap. 8vo, 5s. cloth.
"A valuable contribution to the evidences of revelation, and disposes very conclusively of the arguments of those who would set God's Works against God's Word. No real difficulty is shirked, and no sophistry is left unexposed."—*The Rock.*

Science and Scripture.
SCIENCE ELUCIDATIVE OF SCRIPTURE, AND NOT ANTAGONISTIC TO IT; being a Series of Essays on—1. Alleged Discrepancies; 2. The Theory of the Geologists and Figure of the Earth; 3. The Mosaic Cosmogony; 4. Miracles in general—Views of Hume and Powell; 5. The Miracle of Joshua—Views of Dr. Colenso, &c. By Prof. J. R. Young. Fcap. 5s. cloth.

Geology.
A CLASS-BOOK OF GEOLOGY: Consisting of "Physical Geology," which sets forth the Leading Principles of the Science; and "Historical Geology," which treats of the Mineral and Organic Conditions of the Earth at each successive epoch, especial reference being made to the British Series of Rocks. By Ralph Tate. With more than 250 Illustrations. Fcap. 8vo, 5s. cloth.

Practical Philosophy.
A SYNOPSIS OF PRACTICAL PHILOSOPHY. By Rev. John Carr, M.A., late Fellow of Trin. Coll., Camb. 18mo, 5s. cl.

Pictures and Painters.
THE PICTURE AMATEUR'S HANDBOOK AND DICTIONARY OF PAINTERS: A Guide for Visitors to Picture Galleries, and for Art-Students, including methods of Painting, Cleaning, Re-Lining, and Restoring, Principal Schools of Painting, Copyists and Imitators. By Philippe Daryl, B.A. Cr. 8vo, 3s. 6d. cl.

PUBLISHED BY CROSBY LOCKWOOD & CO.   27

Clocks, Watches, and Bells.
RUDIMENTARY TREATISE on CLOCKS, and WATCHES, and BELLS. By Sir EDMUND BECKETT, Bart. (late E. B. Denison), LL.D., Q.C., F.R.A.S. Sixth Edition, revised and enlarged. Limp cloth (No. 67, Weale's Series), 4s. 6d.; cloth bds. 5s. 6d.
"The best work on the subject probably extant. The treatise on bells is undoubtedly the best in the language."—*Engineering.*
"The only modern treatise on clock-making."—*Horological Journal.*

The Construction of the Organ.
PRACTICAL ORGAN-BUILDING. By W. E. DICKSON, M.A., Precentor of Ely Cathedral. Second Edition, revised, with Additions. 12mo, 3s. cloth boards. [*Just published.*
"In many respects the book is the best that has yet appeared on the subject. We cordially recommend it." —*English Mechanic.*
"The amateur builder will find in this book all that is necessary to enable him personally to construct a perfect organ with his own hands."—*Academy.*

Brewing.
A HANDBOOK FOR YOUNG BREWERS. By HERBERT EDWARDS WRIGHT, B.A. Crown 8vo, 3s. 6d. cloth.
"A thoroughly scientific treatise in popular language."—*Morning Advertiser.*
"We would particularly recommend teachers of the art to place it in every pupil's hands, and we feel sure its perusal will be attended with advantage."—*Brewer.*

Dye-Wares and Colours.
THE MANUAL of COLOURS and DYE-WARES: their Properties, Applications, Valuation, Impurities, and Sophistications. For the Use of Dyers, Printers, Drysalters, Brokers, &c. By J. W. SLATER. Second Edition. Re-written and Enlarged. Crown 8vo, 7s. 6d. cloth. [*Just published.*

Grammar of Colouring.
A GRAMMAR OF COLOURING, applied to Decorative Painting and the Arts. By GEORGE FIELD. New edition. By ELLIS A. DAVIDSON. 12mo, 3s. 6d. cloth.

Woods and Marbles (Imitation of).
SCHOOL OF PAINTING FOR THE IMITATION OF WOODS AND MARBLES, as Taught and Practised by A. R. and P. VAN DER BURG. With 24 full-size Coloured Plates; also 12 Plain Plates, comprising 154 Figures. Folio, 2l. 12s. 6d. bound.

The Military Sciences.
AIDE-MÉMOIRE to the MILITARY SCIENCES. Framed from Contributions of Officers and others connected with the different Services. Originally edited by a Committee of the Corps of Royal Engineers. 2nd Edition, revised; nearly 350 Engravings and many hundred Woodcuts. 3 vols. royal 8vo, cloth, 4l. 10s.

Field Fortification.
A TREATISE on FIELD FORTIFICATION, the ATTACK of FORTRESSES, MILITARY MINING, and RECONNOITRING. By Colonel I. S. MACAULAY, late Professor of Fortification in the R. M. A., Woolwich. Sixth Edition, crown 8vo, cloth, with separate Atlas of 12 Plates, 12s. complete.

Delamotte's Works on Illumination & Alphabets.

A PRIMER OF THE ART OF ILLUMINATION; for the use of Beginners: with a Rudimentary Treatise on the Art, Practical Directions for its Exercise, and numerous Examples taken from Illuminated MSS., printed in Gold and Colours. By F. DELAMOTTE. Small 4to, 9s. Elegantly bound, cloth antique.

"The examples of ancient MSS. recommended to the student, which, with much good sense, the author chooses from collections accessible to all, are selected with judgment and knowledge, as well as taste."—*Athenæum.*

ORNAMENTAL ALPHABETS, ANCIENT and MEDIÆVAL; from the Eighth Century, with Numerals; including Gothic, Church-Text, German, Italian, Arabesque, Initials, Monograms, Crosses, &c. Collected and engraved by F. DELAMOTTE, and printed in Colours. New and Cheaper Edition. Royal 8vo, oblong, 2s. 6d. ornamental boards.

"For those who insert enamelled sentences round gilded chalices, who blazon shop legends over shop-doors, who letter church walls with pithy sentences from the Decalogue, this book will be useful."—*Athenæum.*

EXAMPLES OF MODERN ALPHABETS, PLAIN and ORNAMENTAL; including German, Old English, Saxon, Italic, Perspective, Greek, Hebrew, Court Hand, Engrossing, Tuscan, Riband, Gothic, Rustic, and Arabesque, &c., &c. Collected and engraved by F. DELAMOTTE, and printed in Colours. New and Cheaper Edition. Royal 8vo, oblong, 2s. 6d. ornamental boards.

"There is comprised in it every possible shape into which the letters of the alphabet and numerals can be formed."—*Standard.*

MEDIÆVAL ALPHABETS AND INITIALS FOR ILLUMINATORS. By F. DELAMOTTE. Containing 21 Plates, and Illuminated Title, printed in Gold and Colours. With an Introduction by J. WILLIS BROOKS. Small 4to, 6s. cloth gilt.

THE EMBROIDERER'S BOOK OF DESIGN; containing Initials, Emblems, Cyphers, Monograms, Ornamental Borders, Ecclesiastical Devices, Mediæval and Modern Alphabets, and National Emblems. Collected and engraved by F. DELAMOTTE, and printed in Colours. Oblong royal 8vo, 1s. 6d. ornamental wrapper.

Wood-Carving.

INSTRUCTIONS in WOOD-CARVING, for Amateurs; with Hints on Design. By A LADY. In emblematic wrapper, handsomely printed, with Ten large Plates, 2s. 6d.

"The handicraft of the wood-carver, so well as a book can impart it, may be learnt from 'A Lady's' publication."—*Athenæum.*

Popular Work on Painting.

PAINTING POPULARLY EXPLAINED; with Historical Sketches of the Progress of the Art. By THOMAS JOHN GULLICK, Painter, and JOHN TIMBS, F.S.A. Fourth Edition, revised and enlarged. With Frontispiece and Vignette. In small 8vo, 5s. 6d. cloth.

*** *This Work has been adopted as a Prize-book in the Schools of Art at South Kensington.*

"Contains a large amount of original matter, agreeably conveyed."—*Builder.*

"Much may be learned, even by those who fancy they do not require to be taught, from the careful perusal of this unpretending but comprehensive treatise."—*Art Journal.*

AGRICULTURE, GARDENING, ETC.

Youatt and Burn's Complete Grazier.
THE COMPLETE GRAZIER, and FARMER'S and CATTLE-BREEDER'S ASSISTANT. A Compendium of Husbandry. By WILLIAM YOUATT, ESQ., V.S. 12th Edition, very considerably enlarged, and brought up to the present requirements of agricultural practice. By ROBERT SCOTT BURN. One large 8vo. volume, 860 pp. with 244 Illustrations. 1*l*. 1*s*. half-bound.
"The standard and text-book, with the farmer and grazier."—*Farmer's Magazine.*
"A treatise which will remain a standard work on the subject as long as British agriculture endures."—*Mark Lane Express.*

History, Structure, and Diseases of Sheep.
SHEEP; THE HISTORY, STRUCTURE, ECONOMY, AND DISEASES OF. By W. C. SPOONER, M.R.V.C., &c. Fourth Edition, with fine engravings, including specimens of New and Improved Breeds. 366 pp., 4*s*. cloth.

Production of Meat.
MEAT PRODUCTION. A Manual for Producers, Distributors, and Consumers of Butchers' Meat. Being a treatise on means of increasing its Home Production. Also treating of the Breeding, Rearing, Fattening, and Slaughtering of Meat-yielding Live Stock; Indications of the Quality, etc. By JOHN EWART. Cr. 8vo, 5*s*. cloth.
"A compact and handy volume on the meat question, which deserves serious and thoughtful consideration at the present time."—*Meat and Provision Trades' Review.*

Donaldson and Burn's Suburban Farming.
SUBURBAN FARMING. A Treatise on the Laying Out and Cultivation of Farms adapted to the produce of Milk, Butter and Cheese, Eggs, Poultry, and Pigs. By the late Professor JOHN DONALDSON. With Additions, Illustrating the more Modern Practice, by R. SCOTT BURN. 12mo, 4*s*. cloth boards.

English Agriculture.
THE FIELDS OF GREAT BRITAIN. A Text-book of Agriculture, adapted to the Syllabus of the Science and Art Department. For Elementary and Advanced Students. By HUGH CLEMENTS (Board of Trade). With an Introduction by H. KAINS-JACKSON. 18mo, 2*s*. 6*d*. cloth.
"A clearly written description of the ordinary routine of English farm-life."—*Land.*
"A carefully written text-book of Agriculture."—*Athenæum.* [*Economist.*
"A most comprehensive volume, giving a mass of information."—*Agricultural*

Modern Farming.
OUTLINES OF MODERN FARMING. By R. SCOTT BURN. Soils, Manures, and Crops—Farming and Farming Economy—Cattle, Sheep, and Horses—Management of the Dairy, Pigs, and Poultry—Utilisation of Town Sewage, Irrigation, &c. New Edition. In 1 vol. 1250 pp., half-bound, profusely illustrated, 12*s*.
"There is sufficient stated within the limits of this treatise to prevent a farmer from going far wrong in any of his operations."—*Observer.*

The Management of Estates.
LANDED ESTATES MANAGEMENT: Treating of the Varieties of Lands, Methods of Farming, Farm Building, Irrigation, Drainage, &c. By R. SCOTT BURN. 12mo, 3s. cloth.
"A complete and comprehensive outline of the duties appertaining to the management of landed estates."—*Journal of Forestry.*

The Management of Farms.
OUTLINES OF FARM MANAGEMENT, and the Organization of Farm Labour. Treating of the General Work of the Farm, Field, and Live Stock, Details of Contract Work, Specialties of Labour, Economical Management of the Farmhouse and Cottage, Domestic Animals, &c. By ROBERT SCOTT BURN. 12mo, 3s.

Management of Estates and Farms.
LANDED ESTATES AND FARM MANAGEMENT. By R. SCOTT BURN. (The above Two Works in One Vol.) 6s.

Hudson's Tables for Land Valuers.
THE LAND VALUER'S BEST ASSISTANT: being Tables, on a very much improved Plan, for Calculating the Value of Estates. With Tables for reducing Scotch, Irish, and Provincial Customary Acres to Statute Measure, &c. By R. HUDSON, C.E. New Edition, royal 32mo, leather, gilt edges, elastic band, 4s.

Ewart's Land Improver's Pocket-Book.
THE LAND IMPROVER'S POCKET-BOOK OF FORMULÆ, TABLES, and MEMORANDA, required in any Computation relating to the Permanent Improvement of Landed Property. By JOHN EWART, Land Surveyor. 32mo, leather, 4s.

Complete Agricultural Surveyor's Pocket-Book.
THE LAND VALUER'S AND LAND IMPROVER'S COMPLETE POCKET-BOOK; consisting of the above two works bound together, leather, gilt edges, with strap, 7s. 6d.
"We consider Hudson's book to be the best ready-reckoner on matters relating to the valuation of land and crops we have ever seen, and its combination with Mr. Ewart's work greatly enhances the value and usefulness of the latter-mentioned.—It is most useful as a manual for reference."—*North of England Farmer.*

Grafting and Budding.
THE ART OF GRAFTING AND BUDDING. By CHARLES BALTET. Translated from the French. With upwards of 180 Illustrations. 12mo, 3s. cloth boards. [*Just published.*

Culture of Fruit Trees.
FRUIT TREES, the Scientific and Profitable Culture of. Including Choice of Trees, Planting, Grafting, Training, Restoration of Unfruitful Trees, &c. From the French of DU BREUIL. Third Edition, revised. With an Introduction by GEORGE GLENNY. 4s. cl.
"The book teaches how to prune and train fruit-trees to perfection."- *Field.*

Potato Culture.
POTATOES, HOW TO GROW AND SHOW THEM: A Practical Guide to the Cultivation and General Treatment of the Potato. By JAMES PINK. With Illustrations. Cr. 8vo, 2s. cl.

Good Gardening.
A PLAIN GUIDE TO GOOD GARDENING; or, How to Grow Vegetables, Fruits, and Flowers. With Practical Notes on Soils, Manures, Seeds, Planting, Laying-out of Gardens and Grounds, &c. By S. WOOD. Third Edition. Cr. 8vo, 5s. cloth.
"A very good book, and one to be highly recommended as a practical guide. The practical directions are excellent."—*Athenæum.*

Gainful Gardening.
MULTUM-IN-PARVO GARDENING; or, How to make One Acre of Land produce £620 a year, by the Cultivation of Fruits and Vegetables; also, How to Grow Flowers in Three Glass Houses, so as to realise £176 per annum clear Profit. By SAMUEL WOOD. 3rd Edition, revised. Cr. 8vo, 2s. cloth.
"We are bound to recommend it as not only suited to the case of the amateur and gentleman's gardener, but to the market grower."—*Gardener's Magazine.*

Gardening for Ladies.
THE LADIES' MULTUM-IN-PARVO FLOWER GARDEN, and Amateur's Complete Guide. By S. WOOD. Cr. 8vo, 3s. 6d.

Bulb Culture.
THE BULB GARDEN, or, How to Cultivate Bulbous and Tuberous-rooted Flowering Plants to Perfection. By SAMUEL WOOD. Coloured Plates. Crown 8vo, 3s. 6d. cloth.

Tree Planting.
THE TREE PLANTER AND PLANT PROPAGATOR: A Practical Manual on the Propagation of Forest Trees, Fruit Trees, Flowering Shrubs, Flowering Plants, Pot Herbs, &c. Numerous Illustrations. By SAMUEL WOOD. 12mo, 2s. 6d. cloth.

Tree Pruning.
THE TREE PRUNER: A Practical Manual on the Pruning of Fruit Trees, their Training and Renovation; also the Pruning of Shrubs, Climbers, &c. By S. WOOD. 12mo, 2s. 6d. cloth.

Tree Planting, Pruning, & Plant Propagation.
THE TREE PLANTER, PROPAGATOR, AND PRUNER. By SAMUEL WOOD, Author of "Good Gardening," &c. Consisting of the above Two Works in One Vol., 5s. half-bound.

Early Fruits, Flowers and Vegetables.
THE FORCING GARDEN : or, How to Grow Early Fruits, Flowers, and Vegetables. With Plans and Estimates for Building Glasshouses, Pits, Frames, &c. By S. WOOD. Crown 8vo, 3s. 6d.

Market Gardening, Etc.
THE KITCHEN AND MARKET GARDEN. By Contributors to "The Garden." Compiled by C. W. SHAW, Editor of "Gardening Illustrated." 12mo, 3s. 6d. cl. bds. [*Just published.*

Kitchen Gardening.
KITCHEN GARDENING MADE EASY. Showing how to prepare and lay out the ground, the best means of cultivating every known Vegetable and Herb, etc. By G. M. F. GLENNY. 12mo, 2s.

'A Complete Epitome of the Laws of this Country.'

EVERY MAN'S OWN LAWYER; a Handy-Book of the Principles of Law and Equity. By A BARRISTER. New Edition. Corrected to the end of last Session. Embracing upwards of 3,500 Statements on Points of Law, Verified by the addition of Notes and References to the Authorities. Crown 8vo, cloth, price 6s. 8d. (saved at every consultation).

COMPRISING THE RIGHTS AND WRONGS OF INDIVIDUALS, MERCANTILE AND COMMERCIAL LAW, CRIMINAL LAW, PARISH LAW, COUNTY COURT LAW, GAME AND FISHERY LAWS, POOR MEN'S LAW, THE LAWS OF BANKRUPTCY—BILLS OF EXCHANGE—CONTRACTS AND AGREEMENTS—COPYRIGHT—DOWER AND DIVORCE—ELECTIONS AND REGISTRATION—INSURANCE—LIBEL AND SLANDER—MORTGAGES—SETTLEMENTS—STOCK EXCHANGE PRACTICE—TRADE MARKS AND PATENTS—TRESPASS, NUISANCES, ETC.—TRANSFER OF LAND, ETC.—WARRANTY—WILLS AND AGREEMENTS, ETC.

Also Law for Landlord and Tenant—Master and Servant—Workmen and Apprentices—Heirs, Devisees, and Legatees—Husband and Wife—Executors and Trustees—Guardian and Ward—Married Women and Infants—Partners and Agents—Lender and Borrower—Debtor and Creditor—Purchaser and Vendor—Companies and Associations—Friendly Societies—Clergymen, Churchwardens—Medical Practitioners, &c.—Bankers—Farmers—Contractors—Stock and Share Brokers—Sportsmen and Gamekeepers—Farriers and Horse-Dealers—Auctioneers, House-Agents—Innkeepers, &c.—Pawnbrokers—Surveyors—Railways and Carriers, &c., &c.

" No Englishman ought to be without this book."—*Engineer.*
" What it professes to be—a complete epitome of the laws of this country, thoroughly intelligible to non-professional readers. The book is a handy one to have in readiness when some knotty point requires ready solution."—*Bell's Life.*
" A useful and concise epitome of the law."—*Law Magazine.*

Auctioneer's Assistant.

THE APPRAISER, AUCTIONEER, BROKER, HOUSE AND ESTATE AGENT, AND VALUER'S POCKET ASSISTANT, for the Valuation for Purchase, Sale, or Renewal of Leases, Annuities, and Reversions, and of property generally; with Prices for Inventories, &c. By JOHN WHEELER, Valuer, &c. Fourth Edition, enlarged, by C. NORRIS. Royal 32mo, cloth, 5s.
" A concise book of reference, containing a clearly-arranged list of prices for inventories, a practical guide to determine the value of furniture, &c."—*Standard.*

Auctioneering.

AUCTIONEERS: THEIR DUTIES AND LIABILITIES. By ROBERT SQUIBBS, Auctioneer. Demy 8vo, 10s. 6d. cloth.
" Every auctioneer and valuer ought to possess a copy of this valuable work."—*Ironmonger.*

House Property.

HANDBOOK OF HOUSE PROPERTY: a Popular and Practical Guide to the Purchase, Mortgage, Tenancy, and Compulsory Sale of Houses and Land; including the Law of Dilapidations and Fixtures, &c. By E. L. TARBUCK. 2nd Edit. 12mo, 3s. 6d. cloth.
" We are glad to be able to recommend it."—*Builder.*
" The advice is thoroughly practical."—*Law Journal.*

Metropolitan Rating.

METROPOLITAN RATING: a Summary of the Appeals heard before the Court of General Assessment Sessions at Westminster, in the years 1871-80 inclusive. Containing a large mass of very valuable information with respect to the Rating of Railways, Gas and Waterworks, Tramways, Wharves, Public Houses, &c. By EDWARD and A. L. RYDE. 8vo, 12s. 6d.

Bradbury, Agnew, & Co., Printers, Whitefriars, London.

www.ingramcontent.com/pod-product-compliance
Lightning Source LLC
Chambersburg PA
CBHW022051230426
43672CB00008B/1133